106 Geometry Problems

From the AwesomeMath Summer Program

Titu Andreescu

Michal Rolínek

Josef Tkadlec

Library of Congress Control Number: 2012920774

ISBN-13:978-0-9799269-4-5 ISBN-10:0-9799269-4-7

9 8 7 6 5 4 3 2 1

www.awesomemath.org

Cover design by Iury Ulzutuev

Preface

This book contains 106 geometry problems used in the AwesomeMath Summer Program to train and test top middle and high-school students from the U.S. and around the world. Just like the camp offers both introductory and advanced courses, this book also builds up the material gradually. We begin with a theoretical chapter where we familiarize the reader with basic facts and problem-solving techniques. Then we proceed to the main part of the work, the problem sections.

The problems are a carefully selected and balanced mix which offers a vast variety of flavors and difficulties, ranging from AMC and AIME levels to high-end IMO problems. Out of thousands of Olympiad problems from around the globe we chose those which best illustrate the featured techniques and their applications. The problems meet our demanding taste and fully exhibit the enchanting beauty of classical geometry. For every problem we provide a detailed solution and strive to pass on the intuition and motivation lying behind. Many problems have multiple solutions.

Directly experiencing Olympiad geometry both as contestants and instructors, we are convinced that a neat diagram is essential to efficiently solving a geometry problem. Our diagrams do not contain anything superfluous, yet emphasize the key elements and benefit from a good choice of orientation. Many of the proofs should be legible only from looking at the diagrams.

In the theoretical part we cover the basic theorems concerning circles and ratios and conclude with a short excursion to geometric inequalities. However, we feel that most important are the underlying themes that emphasize the unique combination of Eastern European synthetic feel for geometry and the American more computational approach.

True geometric mastery lies in proficient use of common sense methods, therefore we chose to avoid analytical and computational techniques such as complex numbers, vectors, or barycentric coordinates. A whole new set of topics will be presented in the sequel to this book: *107 Geometry Problems from the AwesomeMath Year-Round Program.*

Although the primary audience for this book consists of high-performing students and their teachers, anyone with an interest in Euclidean geometry or recreational mathematics is invited to join this geometric excursion.

Finally, we would like to express our gratitude to Richard Stong and Cosmin Pohoaţă for critiquing the entire manuscript and providing fruitful comments.

We wish you a pleasant reading.

The Authors

Abbreviations and Notation

Notation of geometrical elements

$\angle BAC$	convex angle by vertex A
$\angle(p, q)$	directed angle between lines p and q
$\angle BAC \equiv \angle B'AC'$	angles BAC and $B'AC'$ coincide
AB	line through points A and B, distance between points A and B
$\overline{AB}$	directed segment from point A to point B
$X \in AB$	X lies on the line AB
$X = AC \cap BD$	X is the intersection of the lines AC and BD
$\triangle ABC$	triangle ABC
$[ABC]$	area of $\triangle ABC$
$[A_1 \dots A_n]$	area of polygon $A_1 \dots A_n$
$AB \parallel CD$	lines AB and CD are parallel
$AB \perp CD$	lines AB and CD are perpendicular
$p(X, \omega)$	power of point X with respect to circle ω
$\triangle ABC \cong \triangle DEF$	triangles ABC and DEF are congruent (in this order of vertices)
$\triangle ABC \sim \triangle DEF$	triangles ABC and DEF are similar (in this order of vertices)

Notation of triangle elements

a, b, c	sides or side lengths of $\triangle ABC$
$\angle A, \angle B, \angle C$	angles by vertices A, B, and C of $\triangle ABC$
s	semiperimeter
x, y, z	expressions $\frac{1}{2}(b+c-a)$, $\frac{1}{2}(c+a-b)$, $\frac{1}{2}(a+b-c)$
r	inradius
R	circumradius
K	area
h_a, h_b, h_c	altitudes in $\triangle ABC$
m_a, m_b, m_c	medians in $\triangle ABC$
l_a, l_b, l_c	angle bisectors (segments) in $\triangle ABC$
r_a, r_b, r_c	exradii in $\triangle ABC$

Abbreviations

AMC10	American Mathematics Contest 10
AMC12	American Mathematics Contest 12
AIME	American Invitational Mathematics Examination
USAJMO	United States of America Junior Mathematical Olympiad
USAMO	United States of America Mathematical Olympiad
USA TST	United States of America IMO Team Selection Test
MEMO	Middle European Mathematical Olympiad
IMO	International Mathematical Olympiad

Contents

Chapter 1

Foundations of Geometry

Preliminaries

We begin our voyage to the fascinating world of classical geometry by reviewing some elementary facts.

Basic Angles

We state the following:

- Vertical angles are equal.
- A line subtends the same angle with any two parallel lines. In other words, alternate angles are equal.
- In triangle ABC we have $AB = AC$ if and only if $\angle B = \angle C$.

The last two parts of the previous statement need to be taken seriously. The second one offers an efficient way to deal with parallel lines and the third one is one of the very few which translates angles into distances and vice versa.

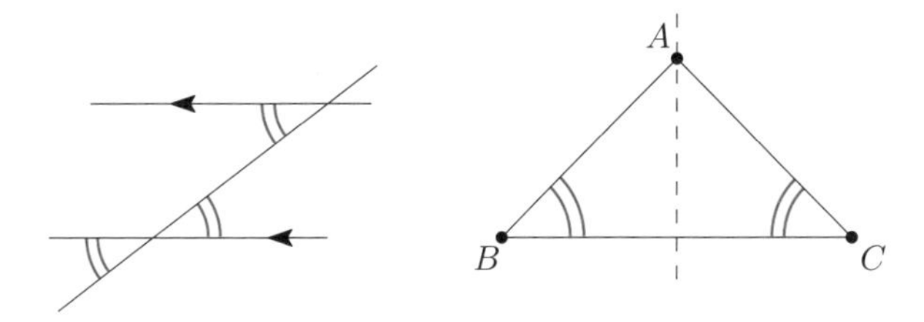

We are ready to prove the universally known theorem on the sum of internal angles in a triangle. In addition, we prove a slight extension, which often offers tiny but pleasant shortcuts in angle calculations.

Proposition 1.1. *Let ABC be a triangle with angles $\angle A$, $\angle B$, $\angle C$. Then:*

(a) $\angle A + \angle B + \angle C = 180^\circ$.
(b) The external angle by vertex C equals $\angle A + \angle B$.

Proof. For (a) draw a line through point A parallel with BC. Since the three angles by vertex A add up to 180°, we arrive at the result by using alternate angles.

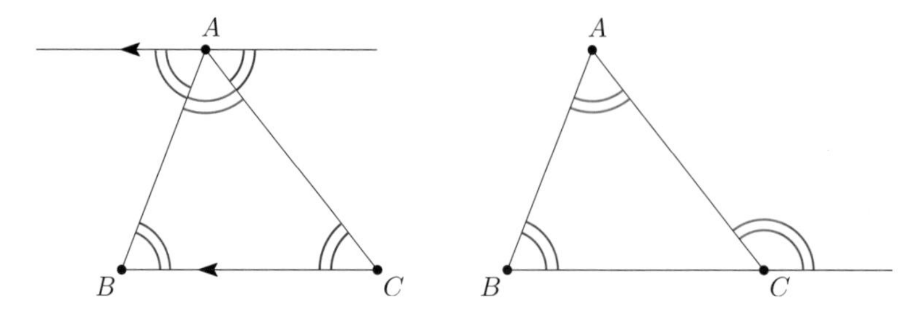

In order to prove (b) we just note that the external angle by vertex C is supplementary to $\angle C$ as well as the sum $\angle A + \angle B$ (by part (a)). □

Also, we know all it takes to prove the Inscribed Angle Theorem, which will later form our understanding of circles.

Theorem 1.2 (Inscribed Angle Theorem). *Let BC be a chord of a circle ω centered at O and let $A \in \omega$, $A \neq B, C$. Then the inscribed angle BAC corresponding to arc BC equals one half of the central angle corresponding to the same arc.*

Proof. Assume first O lies inside triangle ABC.

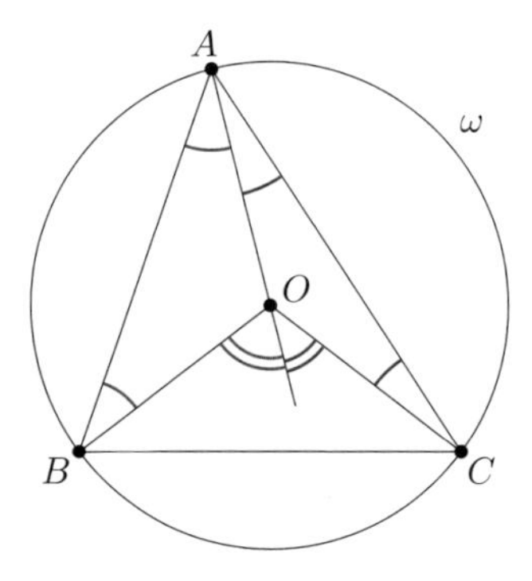

From isosceles triangles OAB and OAC (radii are equal!) we infer $\angle OAB = \angle OBA$ and $\angle OAC = \angle OCA$. Then if we extend ray AO beyond O we can find $\angle BOC$ as sum of two external angles. We see that

$$\angle BOC = 2\angle BAO + 2\angle OAC = 2\angle BAC$$

which is exactly what we wanted.

The case when O lies outside or on the boundary of triangle ABC is treated in the same fashion with a few of the additions becoming subtractions. □

Triangle Congruence and Similarity

Informally, we say that two triangles are congruent if they have the same shape and size. Of course, once two triangles are congruent, their corresponding parts (sides, angles, altitudes, ...) are equal. For proving congruence, we have the following criteria:

- (SSS criterion) If three pairs of sides of two triangles are equal in length, then the triangles are congruent.
- (SAS criterion) If two pairs of sides of two triangles are equal in length, and the included angles are equal, then the triangles are congruent.
- (ASA criterion) If two pairs of angles of two triangles are equal, and two corresponding sides are equal in length, then the triangles are congruent.

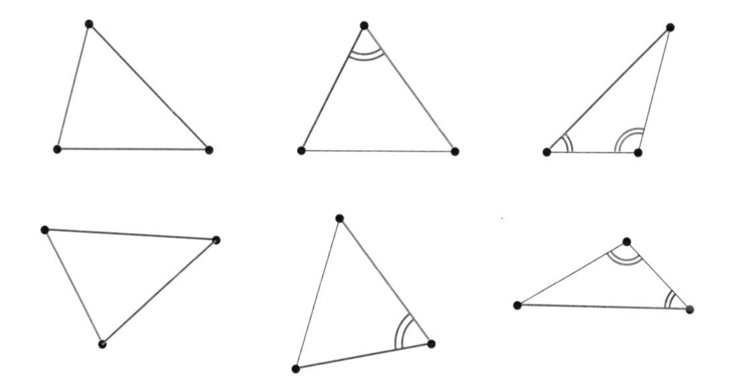

And finally, one criterion which was designed especially for right triangles.

- (HL criterion) If two right triangles have equal hypotenuses and one pair of equal legs, then they are congruent.

For similarity, it is enough for two triangles to have the same shape (i.e. internal angles). Again, similarity implies that all elements of one triangle are just scaled versions of the same elements of the other triangle. Therefore, the ratio of lengths of any corresponding segments is constant. Namely, it is the factor of similarity.

The similarity criteria are the following:

- (AA criterion) If two angles of one triangle are congruent to two angles of another triangle, then the triangles are similar.
- (SAS criterion) If an angle of one triangle is congruent to the corresponding angle of another triangle and the sides that include this angle are proportional, then the two triangles are similar.

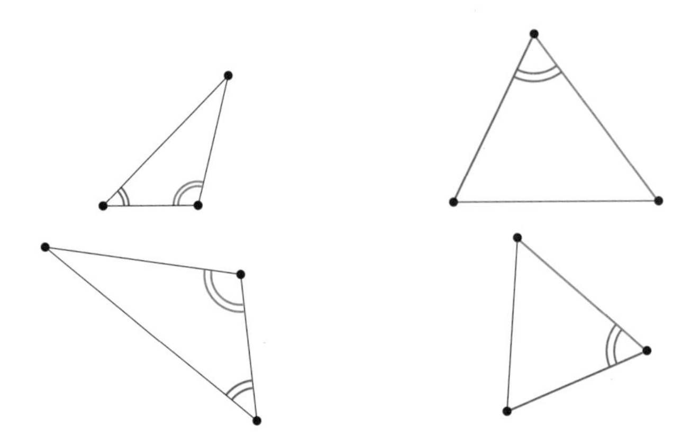

Congruence is most frequently used to give rigorous proofs for very natural claims. Here we prove that a line of symmetry of a segment or an angle indeed has the expected property of being the locus of equidistant points.

Proposition 1.3. *Let A and B be distinct points in the plane. Then the locus of points X for which $XA = XB$ is precisely the perpendicular bisector of AB.*

Proof. Denote by M the midpoint of AB (which is obviously the only satisfying point on AB) and by ℓ the perpendicular bisector of AB.

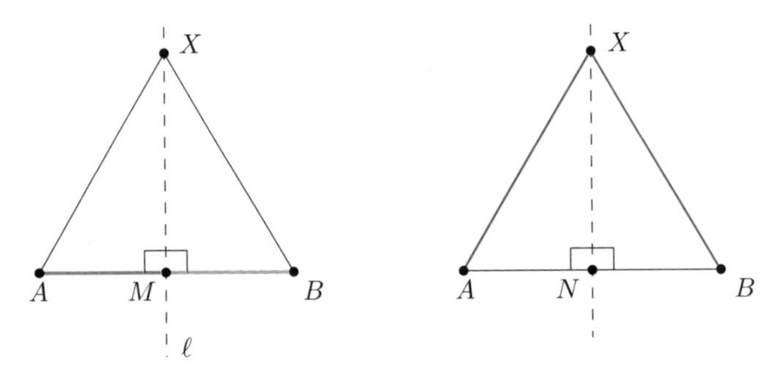

Now if $X \in \ell$ the right triangles AMX and BMX are congruent (SAS: $\angle AMX = \angle BMX = 90°$, $AM = BM$, and XM they have in common) and so $AX = BX$.

On the other hand if $AX = BX$, then let N be the foot of perpendicular from X to AB. Now $\triangle ANX \cong \triangle BNX$ (HL) and thus $AN = NB$ which implies $X \in \ell$. □

Proposition 1.4. *Rays AU and AV form an angle. The locus of points X which have the same distance from the rays AU and AV and lie inside angle UAV is precisely the bisector of $\angle UAV$.*

Proof. Let D and E be the projections of X onto AU and AV, respectively, and let ℓ be the bisector of angle UAV. If $X \in \ell$, then $\triangle ADX \cong \triangle AEX$ (ASA), hence $XD = XE$.

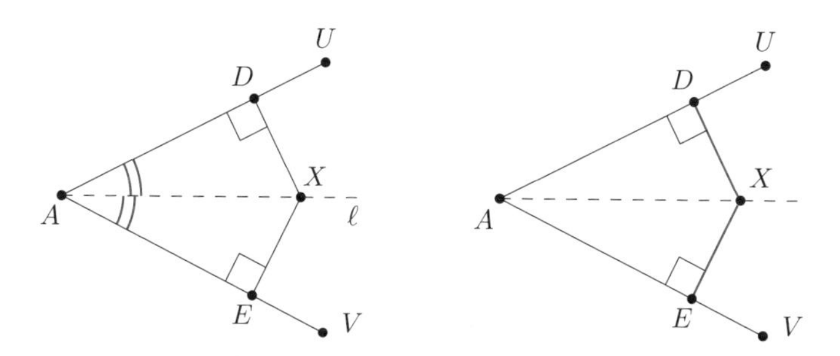

Conversely, if $XD = XE$, we have $\triangle ADX \cong \triangle AEX$ (HL), from which it follows that $\angle XAD = \angle XAE$ and so $X \in \ell$. □

Unlike congruence, similarity has much more striking applications. One of them is that medians divide each other in the ratio $2 : 1$.

Proposition 1.5. *Let ABC be a triangle and let E and F be the midpoints of the sides AB and AC, respectively. Denote by G the intersection of BF and CE. Then $BG = 2GF$ and $CG = 2GE$.*

Proof. First, observe that $\triangle AEF \sim \triangle ABC$ (SAS).

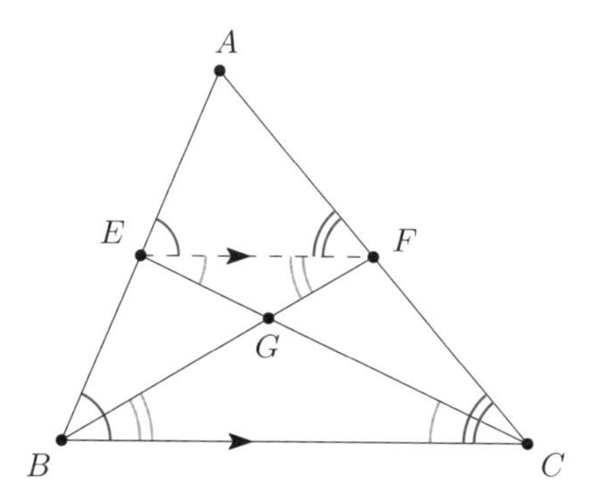

Since the factor of similarity is 2 it follows that $EF = \frac{1}{2}BC$. Moreover, we have $\angle FEA = \angle CBA$, thus $EF \parallel BC$. But then $\angle BCE = \angle CEF$ and we find that $\triangle BCG \sim \triangle FEG$ (AA). Since $EF = \frac{1}{2}BC$, the factor of similarity is $\frac{1}{2}$ and we arrive at the desired equalities $BG = 2GF$ and $CG = 2GE$. □

First Triangle Centers

Despite being such a simple object, the triangle hides perhaps an infinite number of surprising results, many of which are connected to some of its important points. Those are called triangle centers and nowadays over five thousand of them are recognized. Luckily, in olympiad math, it is usually enough to be acquainted with just a small fraction.

Proposition 1.6 (Existence of the Circumcenter). *In triangle ABC the perpendicular bisectors of AB, BC, and CA meet at a single point. This point is called the circumcenter of triangle ABC, is usually denoted by O, and it is the center of the circumscribed circle (or simply circumcircle).*

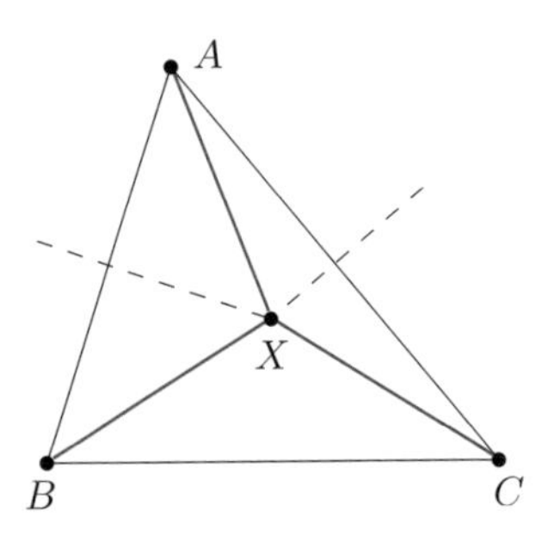

Proof. Let X be the intersection of the perpendicular bisectors of AB and AC. From this we learn $XA = XB$ and $XA = XC$, which gives us $XB = XC$ and this implies that X lies on the perpendicular bisector of BC (if in doubts, see Proposition 1.3).

We have proved that all perpendicular bisectors pass through X. Of course, a circle with center X and radius $XA = XB = XC$ is the circumcircle of triangle ABC. □

Proposition 1.7 (Existence of the Incenter). *In triangle ABC the internal angle bisectors meet at a point. This point is called the incenter of triangle ABC, is usually denoted by I, and it is the center of the incircle of triangle ABC.*

Proof. As expected we denote by X the intersection of the bisectors of $\angle B$ and $\angle C$. Then we know that X is equidistant from the sides AB and BC and also from the sides AC and BC (see Proposition 1.4 if necessary). It follows that is is also equidistant from AB and AC. In other words, it lies on the A-angle bisector. We have found a common point of all three internal angle bisectors.

The circle centered at X having for its radius the common distance from X to the lines BC, CA, and AB is then the incircle of triangle ABC. □

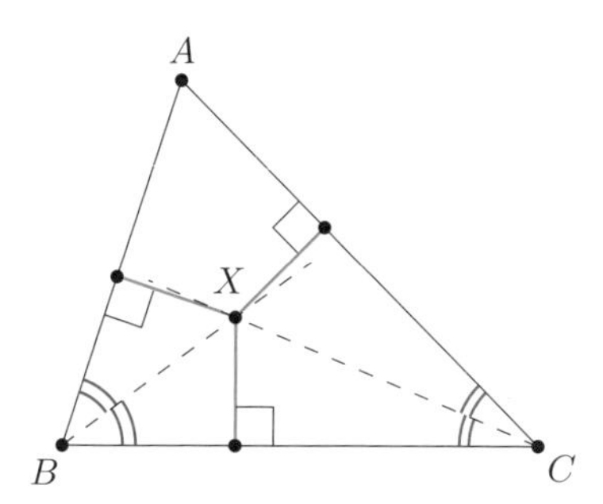

Proposition 1.8 (Existence of the Orthocenter). *In triangle ABC the altitudes meet at a single point. This point is called the orthocenter of triangle ABC and is usually denoted by H.*

Proof. This is a bit tricky! Draw lines parallel with BC, CA, and AB through A, B, and C, respectively and denote the triangle they form by $A'B'C'$ (with $A'B' \parallel AB$ and likewise for the others). The triangles ABC, $A'CB$, $CB'A$, and BAC' are then all congruent (SAS) and in particular A is the midpoint of $B'C'$ and symmetrically for B and C. This implies that the A-altitude in triangle ABC coincides with the perpendicular bisector of $B'C'$ (both of them being perpendicular to $B'C' \parallel BC$ and passing through A). Since the perpendicular bisectors in triangle $A'B'C'$ are concurrent (see Proposition 1.6), so are the altitudes in triangle ABC.

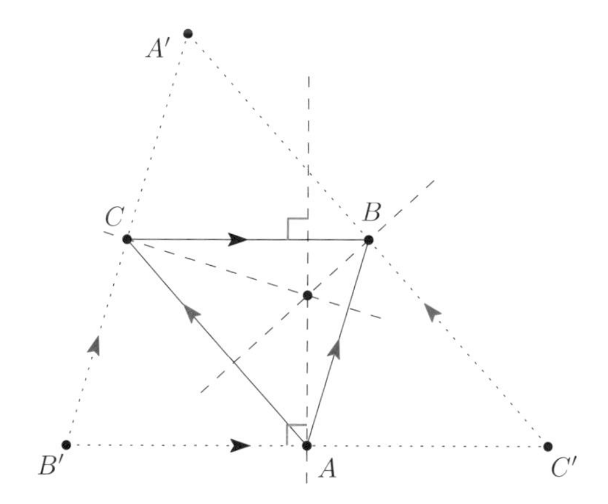

□

Another triplet of circles bound to a triangle are the excircles. They are in many ways analogous to the incircle and posses numerous remarkable properties.

Proposition 1.9 (Existence of the Excenter). *In triangle ABC the A-angle bisector and the bisectors of external angle B and C meet at a point. This*

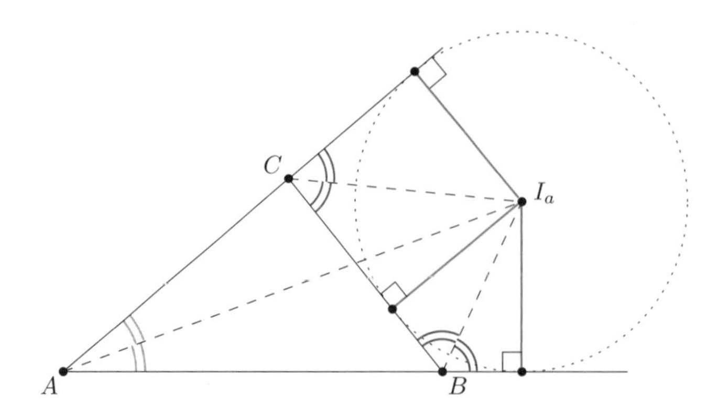

point is called the A-excenter of triangle ABC, is usually denoted by I_a and it is the center of the A-excircle (circle tangent to the side BC and to the extended sidelines AB and AC). Similarly, we define points I_b and I_c.

Proof. Do it yourself! □

Proposition 1.10 (Existence of the Centroid). *In triangle ABC the medians meet at a point. This point is called the centroid of triangle ABC and is usually denoted by G.*

Proof. Let X be the intersection of the median AM with the B-median and let X' be the intersection of AM with the C-median. From Proposition 1.5 we know that both $AX = 2XM$ and $AX' = 2X'M$, thus inevitably $X = X'$ and the concurrence is proved.

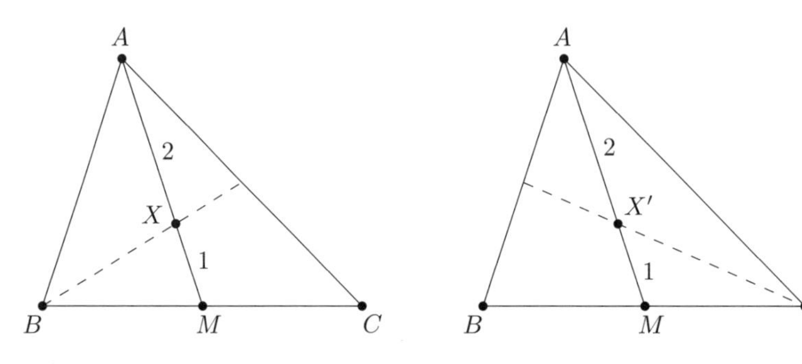

□

One may be tempted to believe the following result is not worth remembering as it is so easy to derive. But this would be a terrible mistake! In fact, many connections are revealed if one knows this without thinking!

Proposition 1.11. *Let ABC be a triangle with orthocenter H, incenter I, and circumcenter O. Then:*

(a) If triangle ABC is acute, then $\angle BHC = 180° - \angle A$.

(b) $\angle BIC = 90^\circ + \frac{1}{2}\angle A$.
(c) If $\angle A$ *is acute, then* $\angle BOC = 2\angle A$.

Proof. In (a) we denote by B' and C' the feet of the altitudes from B and C, respectively, and focus on quadrilateral B_0HC_0A.

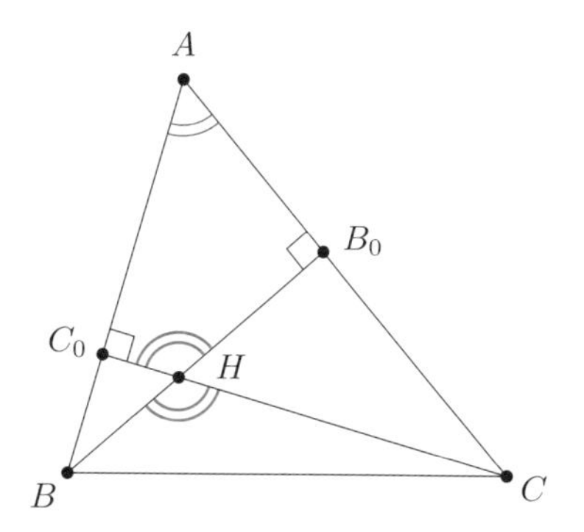

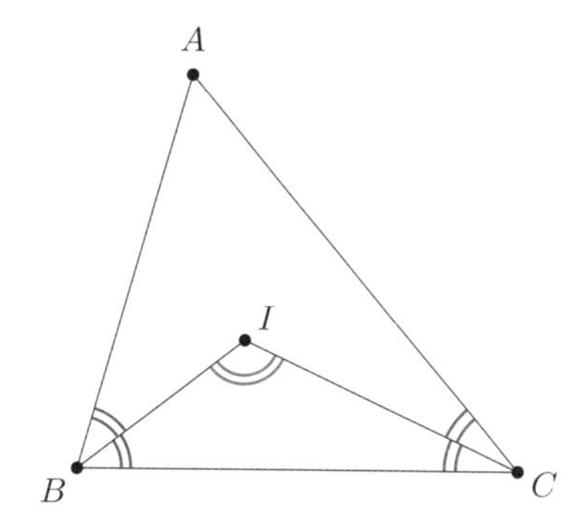

Since the sum of internal angles in a quadrilateral is 360° and $\angle HB_0A = \angle HC_0A = 90^\circ$, the remaining two angles add up to 180°. In other words:

$$\angle BHC = \angle C_0HB_0 = 180^\circ - \angle A.$$

For (b) we angle-chase in triangle BIC. Since BI and CI are angle bisectors, we have

$$\angle BIC = 180^\circ - \frac{1}{2}\angle B - \frac{1}{2}\angle C = 90^\circ + \left(90^\circ - \frac{1}{2}\angle B - \frac{1}{2}\angle C\right) = 90^\circ + \frac{1}{2}\angle A.$$

Finally, (c) is just a consequence of the Inscribed Angle Theorem. □

Metric Relations

Equal Tangents

We start with a very simple method, which on the other hand has many nontrivial applications and appears over and over again in contests. We will just play with equal segments.

Proposition 1.12 (Equal Tangents). *Two tangent lines to the given circle ω intersect at A. Denote by B, C the points of tangency with the circle. Then $AB = AC$.*

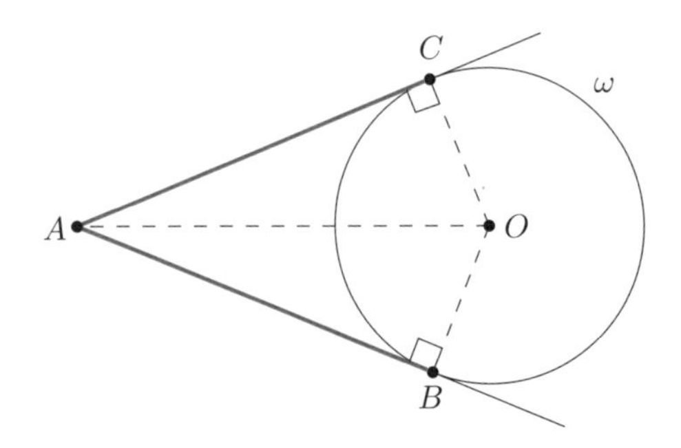

Proof. Let O be the center of ω. Then $\angle OBA = 90^\circ = \angle OCA$ and $OB = OC$. Moreover, the right triangles OAB and OAC share the hypotenuse OA, thus they are congruent (HL). The result follows. □

Proposition 1.13. *Let p, q be common external tangents of circles ω_1, ω_2. Denote by A, B the points of tangency of p with ω_1 and ω_2, respectively, and by C, D the points of tangency of q with ω_1 and ω_2, respectively. Then:*

(a) $AB = CD$.

(b) If the two circles are nonintersecting and their common internal tangent r intersects p and q at points X, Y, respectively, we have $AB = CD = XY$.

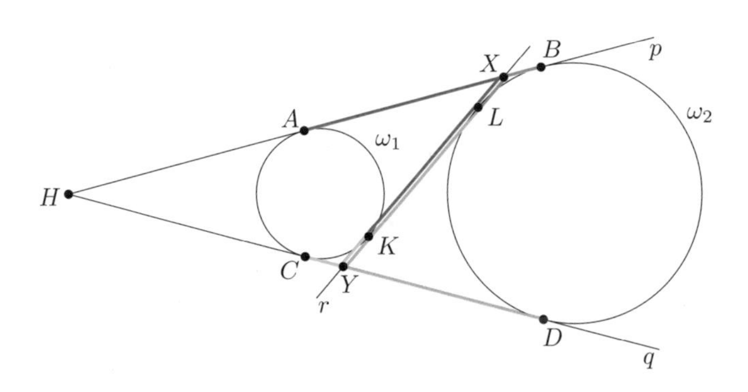

Proof. For part (a), if $AB \parallel CD$ then $ABCD$ is a rectangle and we are done. Otherwise, let $AB \cap CD = H$. Now by Equal Tangents we have $HA = HC$ and $HB = HD$. Subtracting gives the result.

In part (b), denote by K, L the points of tangency of r and ω_1, ω_2, respectively. Now using part (a) and Equal Tangents several times, we obtain

$$2 \cdot XY = (XL + YL) + (YK + XK) = \\ = XB + YD + YC + XA = AB + CD = 2 \cdot AB.$$

□

Theorem 1.14 (Pitot[1] Theorem). *Let $ABCD$ be a quadrilateral with an inscribed circle. Then*

$$AB + CD = BC + DA.$$

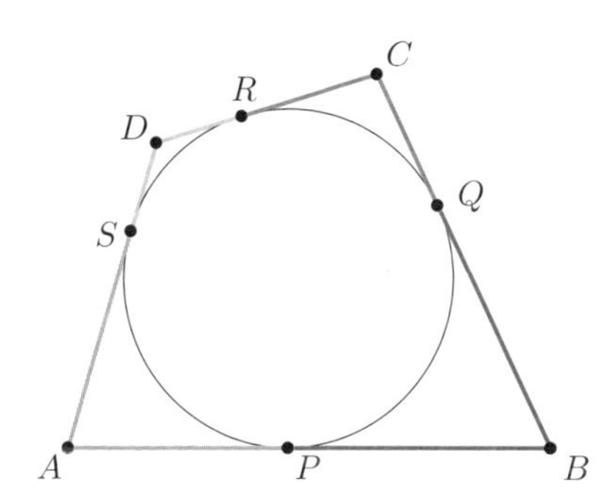

Proof. Denote by P, Q, R, S the points of tangency of the inscribed circle with the sides AB, BC, CD, DA, respectively. Equal Tangents give

$$AB + CD = AP + BP + CR + DR = AS + BQ + CQ + DS = BC + DA$$

and we are done. □

In fact, the condition $AB + CD = BC + DA$ is also sufficient for a quadrilateral to have an inscribed circle. Can you find the proof?

We are ready to prove an important fact from the geometry of a triangle.

Proposition 1.15. *Let ABC be a triangle with semiperimeter s. Denote by D, E, F the points of tangency of the incircle with the sides BC, CA, AB, respectively. Also let the A-excircle touch the lines BC, CA, AB at points K, L, M, respectively. Then the following holds:*

(a) $2 \cdot AE = 2 \cdot AF = -a + b + c$, $\quad 2 \cdot BD = 2 \cdot BF = a - b + c$, $\quad 2 \cdot CD = 2 \cdot CE = a + b - c$.

[1]Henri Pitot (1695–1771) was a French hydraulic engineer.

(b) $2AL = 2AM = a + b + c$, *in other words* $AL = AM = s$.
(c) Points K *and* D *are symmetric with respect to the midpoint of* BC.

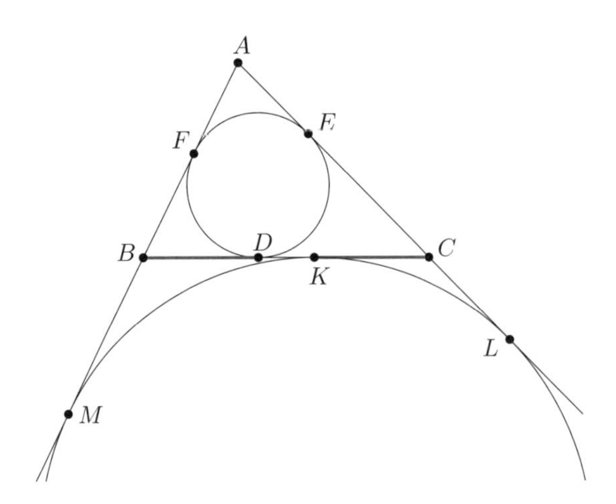

Proof. For part (a), we again make use of Equal Tangents. Namely

$$2 \cdot AE = AE + AF = (b - CE) + (c - BF) = b + c - (CD + BD) = b + c - a.$$

The other two relations are proved analogously.
Similarly, we obtain part (b):

$$2 \cdot AL = AL + AM = (b + CL) + (c + BM) = b + c + (CK + BK) = a + b + c.$$

To prove part (c), it suffices to show that $BD = CK$ and indeed using parts (a) and (b), we easily derive

$$2 \cdot BD = a - b + c = 2(s - b) = 2(AL - AC) = 2 \cdot CL = 2 \cdot CK$$

and thus conclude the proof. □

This result immediately gives a convenient formula for the inradius of a right triangle.

Proposition 1.16. *In right triangle* ABC, *where* $\angle A = 90°$, *denote by* r *the radius of its incircle* ω. *Then*

$$r = \frac{AB + AC - CB}{2}.$$

Proof. Denote by E and F the points of contact of ω with AC and AB, respectively, and by I the center of ω.

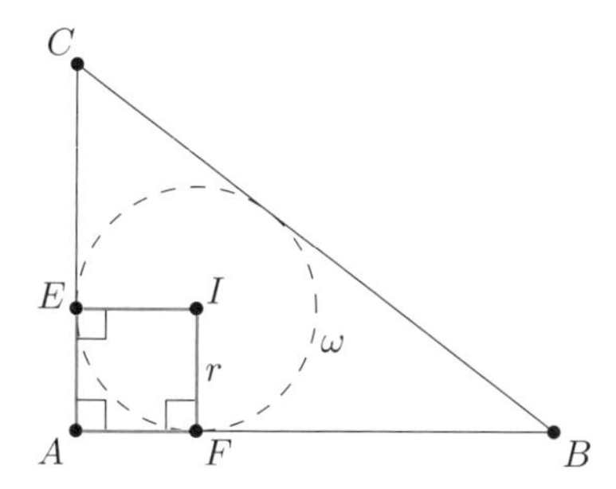

We take a look at quadrilateral $AFIE$. It has three right angles and equal pairs of adjacent sides $AF = AE$ (Equal Tangents) and $IE = IF$, which makes it a square. Thus $r = AE$ and the previous proposition gives the desired

$$r = AE = \frac{AB + AC - CB}{2}.$$

□

Example 1.1. *Let $ABCD$ be a parallelogram with $AB > BC$. Let K, M be the points of tangency of the incircles of triangles ACD and ABC with AC, respectively. Similarly, let L, N be the points of tangency of the incircles of triangles BCD and ABD with BD, respectively. Prove that $KLMN$ is a rectangle.*

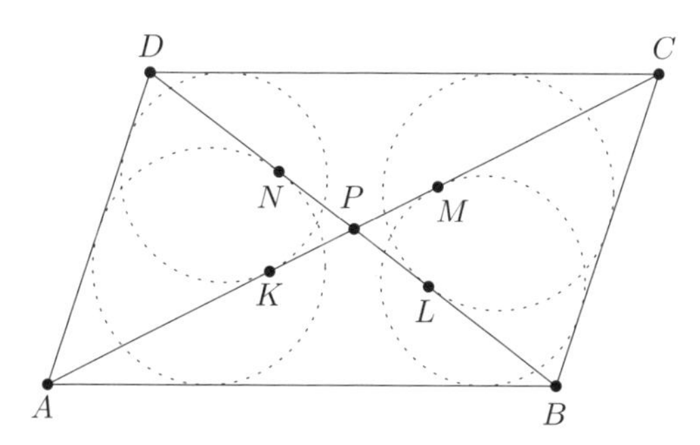

Proof. Let P be the intersection of diagonals in $ABCD$. First, we observe that triangles ABD and CDB are congruent (ASA) and symmetric with respect to P. Therefore, we have $PN = PL$. Similarly, we deduce that $PK = PM$, thus the diagonals in $KLMN$ bisect each other and this quadrilateral is a parallelogram.

To prove that it is in fact a rectangle, it suffices to show $NL = KM$ or equivalently $PN = PK$. By Proposition 1.15(a) applied to triangle ABD we get

$$PN = \frac{DB}{2} - DN = \frac{DB}{2} - \frac{DB + DA - AB}{2} = \frac{AB - DA}{2}.$$

Analogously, we calculate the length of PK:

$$PK = \frac{AC}{2} - AK = \frac{AC}{2} - \frac{AC + DA - CD}{2} = \frac{CD - DA}{2}.$$

Since $AB = CD$, the conclusion follows. □

The Law of Sines

Now we will discuss one of the most fundamental theorems from the triangle geometry, the Law of Sines. In fact, the method of trigonometric computation is a very powerful technique and for an aspiring contestant it is a must to know. One of the reasons why the Law of Sines is so useful is the well-known fact that $\sin x = \sin(180° - x)$.

Theorem 1.17 (The Extended Law of Sines). *Let ABC be a triangle. Then*

$$\frac{a}{\sin \angle A} = \frac{b}{\sin \angle B} = \frac{c}{\sin \angle C} = 2R,$$

where R is the circumradius of triangle ABC.

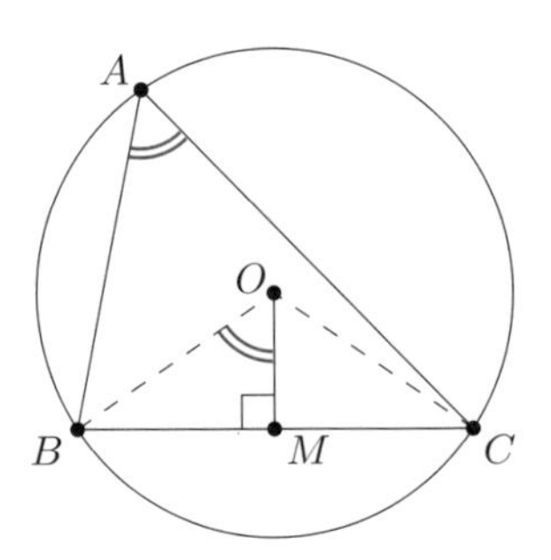

Proof. Assume that $\angle A$ is acute. Let O be the circumcenter of triangle ABC and M the midpoint of BC. Then $\angle BOC = 2\angle A$ as it is a central angle. Triangle OBC is isosceles, so OM bisects $\angle BOC$ and $\angle BOM = \angle A$. From right triangle BOM we obtain

$$\sin \angle A = \frac{\frac{1}{2}a}{R},$$

implying the result. The case when $\angle A$ is not acute can be proved analogously using the fact that $\sin \angle A = \sin(180° - \angle A)$. Details are left for the reader. □

The following lemma provides a useful shortcut whenever we are dealing with ratios in adjacent triangles.

Proposition 1.18 (Ratio Lemma). *In triangle ABC, let $D \in BC$ and denote by α_1 and α_2 the angles BAD and DAC, respectively. Then*

$$\frac{BD}{DC} = \frac{AB \cdot \sin\alpha_1}{AC \cdot \sin\alpha_2}.$$

Proof. Denote the angle ADB by φ. By the Law of Sines in adjacent triangles ADB and ADC, we obtain

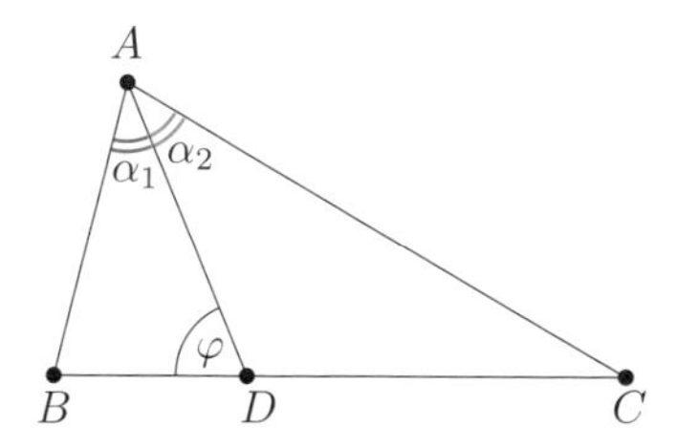

$$\frac{BD}{\sin\alpha_1} = \frac{AB}{\sin\varphi}, \qquad \frac{CD}{\sin\alpha_2} = \frac{AC}{\sin(180^\circ - \varphi)} = \frac{AC}{\sin\varphi}.$$

Now dividing the two relations yields the result. □

As an immediate corollary we obtain the well-known Angle Bisector Theorem.

Theorem 1.19 (Angle Bisector Theorem). *In triangle ABC let AD, $D \in BC$, be the internal angle bisector. Then*

$$\frac{BD}{CD} = \frac{c}{b}, \qquad BD = \frac{ac}{b+c}, \qquad CD = \frac{ab}{b+c}.$$

Proof. The first relation comes directly from the Ratio Lemma as $\alpha_1 = \alpha_2$. The other two are the solutions to system of equations

$$\frac{BD}{CD} = \frac{c}{b} \quad \text{and} \quad BD + CD = a.$$

□

The next example illustrates typical use of the Law of Sines.

Example 1.2 (Germany 2003). *Let $ABCD$ be a parallelogram. Let X and Y be points on the sides AB and BC, respectively, such that $AX = CY$. Prove that the intersection of lines AY and CX lies on the angle bisector of $\angle ADC$.*

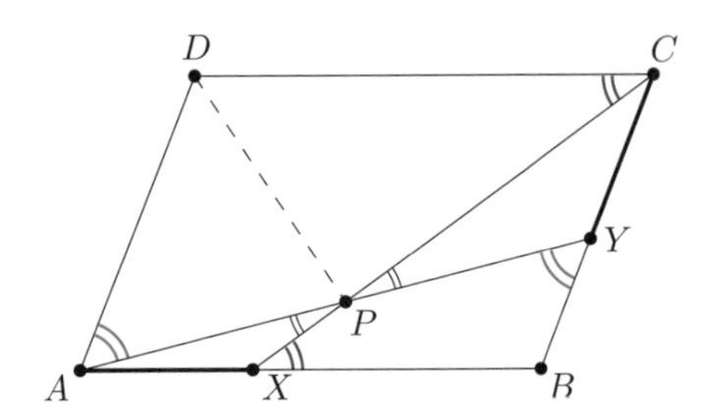

Proof. Denote the intersection of AY and CX by P. First, we make use of parallel lines to see that $\angle DAP = 180^\circ - \angle PYC$ and $\angle DCP = 180^\circ - \angle PXA$. Now we are ready to use the Law of Sines successively in triangles APD and APX to obtain

$$\sin\angle ADP = \frac{\sin\angle DAP}{PD}\cdot AP = \frac{\sin\angle DAP}{PD}\cdot AX\cdot\frac{\sin\angle PXA}{\sin\angle APX},$$

and similarly

$$\sin\angle CDP = \frac{\sin\angle DCP}{PD}\cdot CP = \frac{\sin\angle DCP}{PD}\cdot CY\cdot\frac{\sin\angle PYC}{\sin\angle CPY}.$$

But $AX = CY$, $\angle APX = \angle CPY$, and we already know that $\sin\angle DAP = \sin\angle PYC$ and $\sin\angle DCP = \sin\angle PXA$, hence $\sin\angle ADP = \sin\angle CDP$. Moreover, $\angle ADP + \angle CDP \neq 180^\circ$, thus $\angle ADP = \angle CDP$. □

The Law of Cosines

We present another theorem from basic triangle geometry, the Law of Cosines. Again this theorem is more useful than it might seem at the first glance.

Theorem 1.20 (Law of Cosines)**.** *Let ABC be a triangle. Then*

$$a^2 = b^2 + c^2 - 2bc\cos\angle A.$$

Proof. Assume triangle ABC is acute. Let AK, BL, CM be the altitudes in triangle ABC. Using right triangles yields

$$\begin{aligned} a^2 &= a(BK + KC) = a(c\cos\angle B) + a(b\cos\angle C) = \\ &= c(a\cos\angle B) + b(a\cos\angle C) = \\ &= c\cdot BM + b\cdot CL = c(c - b\cos\angle A) + b(b - c\cos\angle A) = \\ &= c^2 + b^2 - 2bc\cos\angle A. \end{aligned}$$

The cases when triangle ABC is right or obtuse can be done analogously and we leave them as an exercise for the reader. □

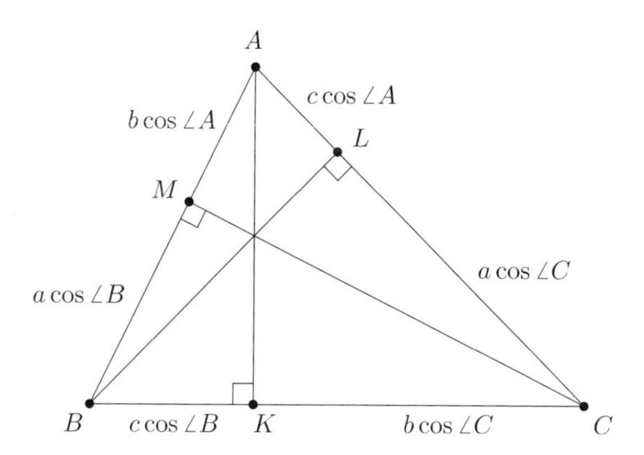

Corollary 1.21 (Generalized Pythagorean[2] Theorem). *Let ABC be a triangle. Then:*

(a) $\angle C < 90^\circ$ *if and only if* $a^2 + b^2 > c^2$.
(b) $\angle C = 90^\circ$ *if and only if* $a^2 + b^2 = c^2$.
(c) $\angle C > 90^\circ$ *if and only if* $a^2 + b^2 < c^2$.

Proof. This is a direct consequence of the Law of Cosines and the fact that the function cos is positive on $(0^\circ, 90^\circ)$, negative on $(90^\circ, 180^\circ)$, and zero for 90°. □

Example 1.3 (USAMO 1996, Titu Andreescu). *Let ABC be a triangle and M an interior point such that* $\angle MAB = 10^\circ$, $\angle MBA = 20^\circ$, $\angle MAC = 40^\circ$, *and* $\angle MCA = 30^\circ$. *Prove that triangle ABC is isosceles.*

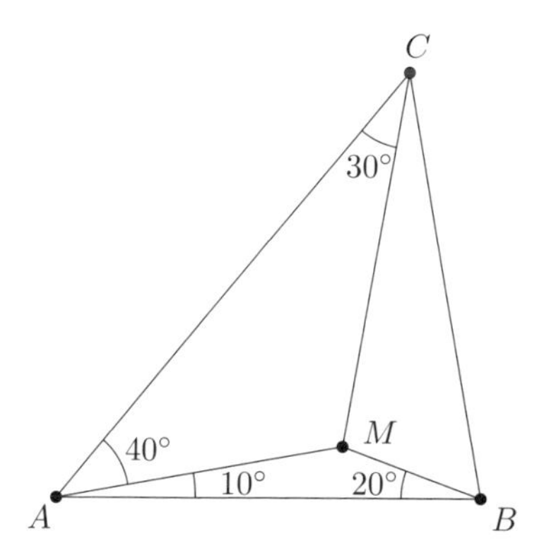

Proof. Assume without loss of generality $AB = 1$. First note $\angle AMB = 150^\circ$ and $\angle AMC = 110^\circ$. The key is to realize that we can calculate all three side

[2]Pythagoras of Samos (c. 570–495 BC) was a greek philosopher and mathematician.

lengths. Indeed, using the Law of Sines in triangles AMB and AMC, and the formula $\sin 2x = 2\sin x \cos x$ yields

$$AM = AB \cdot \frac{\sin 20^\circ}{\sin 150^\circ} = 2\sin 20^\circ$$

and

$$AC = AM \cdot \frac{\sin 110^\circ}{\sin 30^\circ} = (2\sin 20^\circ) \cdot 2 \cdot \cos 20^\circ = 2\sin 40^\circ.$$

From the Law of Cosines in triangle ABC we obtain

$$\begin{aligned} BC^2 &= 1^2 + (2\sin 40^\circ)^2 - 2 \cdot 1 \cdot 2\sin 40^\circ \cos 50^\circ \\ &= 1 + 4\sin^2 40^\circ - 4\sin^2 40^\circ = 1. \end{aligned}$$

Thus $AB = BC$ and triangle ABC is isosceles as desired. □

Now we shall present a metric criterion for orthogonality, that is a straightforward application of the Law of Cosines.

Proposition 1.22. *Let AC and BD be two lines in plane. Then, $AC \perp BD$ if and only if*

$$AB^2 + CD^2 = AD^2 + BC^2.$$

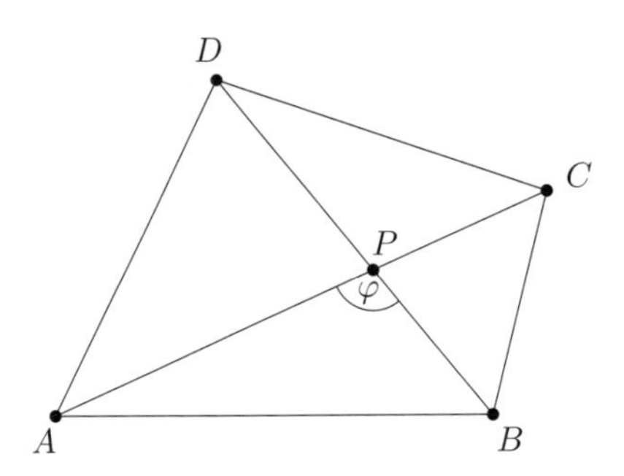

Proof. Assume $ABCD$ is a convex quadrilateral. Denote by P the intersection of diagonals and let $\angle APB = \varphi$. Write the Law of Cosines in triangles ABP and CDP:

$$\begin{aligned} AB^2 &= AP^2 + BP^2 - 2 \cdot AP \cdot BP \cdot \cos\varphi, \\ CD^2 &= CP^2 + DP^2 - 2 \cdot CP \cdot DP \cdot \cos\varphi. \end{aligned}$$

Adding the equations we obtain

$$AB^2 + CD^2 = AP^2 + BP^2 + CP^2 + DP^2 - 2(CP \cdot DP + AP \cdot BP)\cos\varphi.$$

Similarly, we add the Law of Cosines from triangles BCP and DAP, keeping in mind that $\cos x = -\cos(180° - x)$. We get

$$BC^2 + DA^2 = BP^2 + CP^2 + DP^2 + AP^2 + 2(BP \cdot CP + DP \cdot AP)\cos\varphi.$$

Comparing, we see that $AB^2 + CD^2 = BC^2 + DA^2$ holds if and only if

$$(CP \cdot DP + AP \cdot BP + BP \cdot CP + DP \cdot AP)\cos\varphi = 0.$$

Since the former quantity is positive, this may only happen if $\cos\varphi = 0$ i.e. if $\varphi = 90°$.

The case when $ABCD$ is not convex is handled analogously and the cases when $ABCD$ is not a quadrilateral are even easier. □

Theorem 1.23 (Stewart's[3] theorem). *In triangle ABC let D lie on the side BC. Denote the distances BD, DC, and AD by m, n, and d, respectively. Then*

$$a(d^2 + mn) = b^2 m + c^2 n.$$

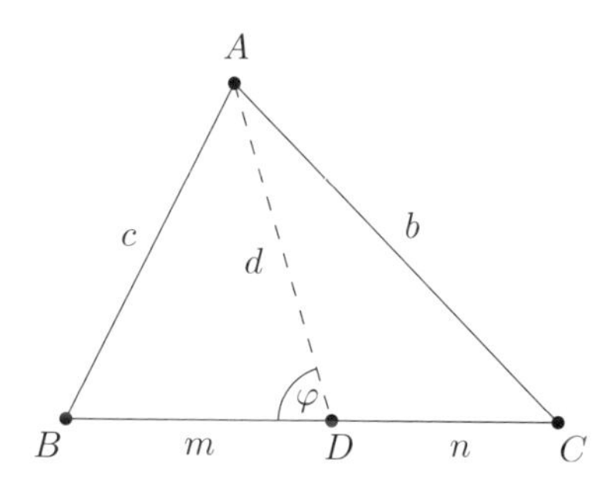

Proof. Denote the angle ADB by φ and write the Law of Cosines in adjacent triangles ABD and ADC in the following way:

$$c^2 = d^2 + m^2 - 2md\cos\varphi, \quad b^2 = d^2 + n^2 + 2nd\cos\varphi.$$

Multiplying the first relation by n and the second by m so that we can easily eliminate the cosine just by adding the equations, we obtain

$$c^2 n + b^2 m = d^2 n + m^2 n + d^2 m + n^2 m = (m+n)(d^2 + mn) = a(d^2 + mn),$$

and we are done. □

[3]Matthew Stewart (1719–1785) was a Scottish mathematician and minister of religion.

Corollary 1.24. *In triangle ABC let M be the midpoint of BC and let AD, $D \in BC$, be the internal angle bisector. Then*

$$m_a^2 = AM^2 = \frac{b^2 + c^2}{2} - \frac{a^2}{4}, \quad l_a^2 = AD^2 = bc\left(1 - \left(\frac{a}{b+c}\right)^2\right).$$

Proof. The first rclation is immediate as we have $m = n = \frac{1}{2}a$ in Stewart's theorem.

For the second part we recall the Angle Bisector Theorem (see Theorem 1.19) and apply Stewart's theorem to obtain

$$a\left(l_a^2 + \frac{a^2bc}{(b+c)^2}\right) = \frac{b^2ac}{b+c} + \frac{c^2ab}{b+c}.$$

After dividing by a and simplifying the right-hand side we get

$$l_a^2 + \frac{a^2bc}{(b+c)^2} = bc,$$

which is easily seen to be equivalent to what we are proving. □

Areas

Now we shall develop some interesting properties of areas, starting only with the basic formula for area of a triangle, namely $2K = ah_a = bh_b = ch_c$. Also we will see how calculating ratios of areas may be helpful. Recall that area of triangle ABC is denoted either by K or by $[ABC]$.

Proposition 1.25. *Let ABC be a triangle. Then we can calculate its area in the following ways:*

(a) $K = \frac{1}{2}ab\sin\angle C = \frac{1}{2}bc\sin\angle A = \frac{1}{2}ca\sin\angle B = abc/(4R)$.
(b) $K = \sqrt{s(s-a)(s-b)(s-c)}$ *(Heron's[4] formula).*
(c) $K = rs = r_a(s-a) = r_b(s-b) = r_c(s-c)$.

Proof. For the first part of (a), it suffices to prove (by symmetry) $2K = ab\sin\angle C$, but as $h_a = b\sin\angle C$ this is obvious. Also, by the Extended Law of Sines

$$\frac{1}{2}ab\sin\angle C = \frac{1}{2}ab\frac{c}{2R} = \frac{abc}{4R}.$$

By part (a) we have $16K^2 = 4a^2b^2\sin^2\angle C = 4a^2b^2(1 - \cos^2\angle C)$ and by the Law of Cosines

$$\cos\angle C = \frac{a^2 + b^2 - c^2}{2ab}.$$

[4]Heron of Alexandria (c. 10–70) was an ancient Greek mathematician and engineer.

Putting this together we obtain

$$16K^2 = 4a^2b^2\left(1 - \frac{(a^2+b^2-c^2)^2}{4a^2b^2}\right) = 4a^2b^2 - (a^2+b^2-c^2)^2,$$

which factors as

$$(2ab - a^2 - b^2 + c^2)(2ab + a^2 + b^2 - c^2) = \left(c^2 - (a-b)^2\right)\left((a+b)^2 - c^2\right) =$$
$$= (-a+b+c)(a-b+c)(a+b-c)(a+b+c),$$

but this is just an equivalent form of Heron's formula so we have proved part (b).

Finally for (c), denote by I the incenter of triangle ABC and observe that

$$K = [BIC] + [CIA] + [AIB] = \frac{1}{2}(ra + rb + rc) = rs.$$

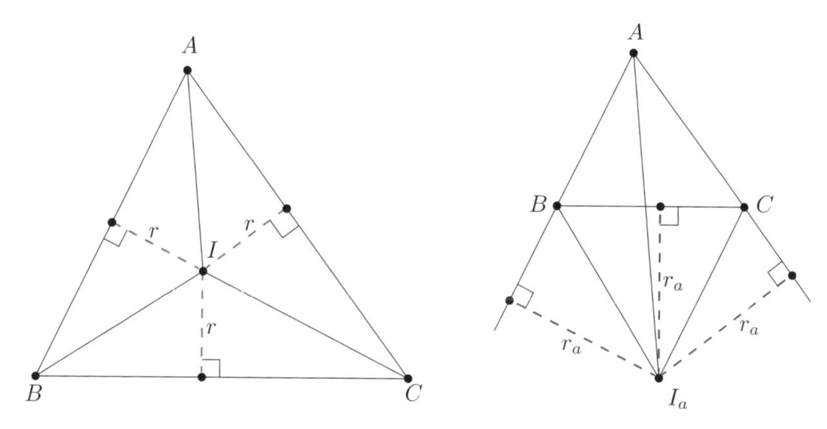

Similarly, denote by I_a the A-excenter of triangle ABC and again observe

$$K = [AI_aC] + [BI_aA] - [BI_aC] = \frac{1}{2}(br_a + cr_a - ar_a) = r_a(s-a).$$

The remaining relations are proved in a similar fashion, so the proof is finished. □

The main application of these area formulas is that they allow us to express some common triangle elements in terms of triangle sides. This is especially convenient if we use the standard notation xyz:

$$x = s - a = \frac{1}{2}(b+c-a), \quad y = s - b = \frac{1}{2}(c+a-b), \quad z = s - c = \frac{1}{2}(a+b-c).$$

These expressions come in fact from Proposition 1.15. Another way to look at x, y, and z is that they are the unique numbers such that

$$a = y + z, \qquad b = z + x, \qquad c = x + y.$$

Since calculating triangle elements is a common theme in many problems, we will appreciate that the xyz notation simplifies the computations.

Proposition 1.26 (xyz formulas). *In triangle ABC we can find the area K, inradius r, and circumradius R in terms of xyz as follows:*

(a)

$$K = \sqrt{(x+y+z)xyz},$$

(b)

$$r = \sqrt{\frac{xyz}{x+y+z}},$$

(c)

$$R = \frac{(y+z)(z+x)(x+y)}{4\sqrt{xyz(x+y+z)}}.$$

Proof. For (a) we just rewrite Heron's formula

$$K = \sqrt{s(s-a)(s-b)(s-c)} = \sqrt{(x+y+z)xyz}.$$

To find the inradius we use the formula $K = rs$, from which we find $r = K/(x+y+z)$ and we conclude by using the result of (a).

Finally, the circumradius R appears in the formula $K = (abc)/(4R)$, thus

$$R = \frac{(y+z)(z+x)(x+y)}{4K}$$

and again part (a) ensures the rest. □

The following lemma is surprisingly handy, because it efficiently transfers ratios of distances to ratios of areas. We will refer to this result as the *Area Lemma.*

Proposition 1.27 (Area Lemma). *Let ABC be a triangle, $D \in BC$ and $X \in AD$. Then*

$$\frac{[BCX]}{[BCA]} = \frac{DX}{DA}.$$

Proof. Drop perpendiculars from A and X to BC and denote their feet by A_0 and X_0, respectively. The triangles DX_0X and DA_0A are similar (AA). Hence

$$\frac{[BCX]}{[BCA]} = \frac{\frac{1}{2} \cdot BC \cdot XX_0}{\frac{1}{2} \cdot BC \cdot AA_0} = \frac{XX_0}{AA_0} = \frac{DX}{DA}.$$

□

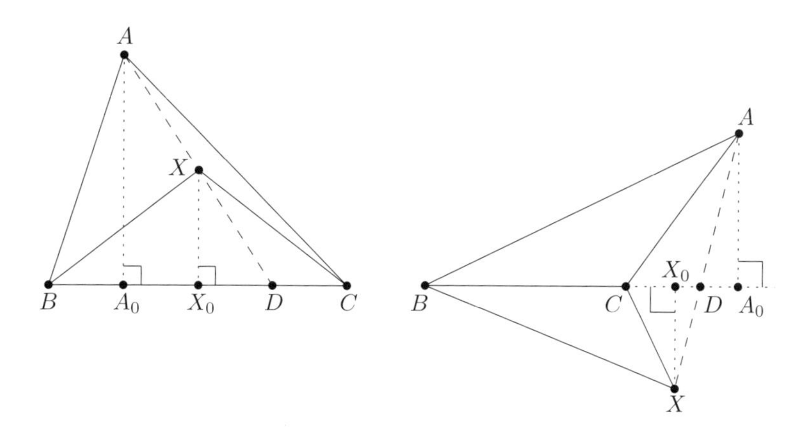

Example 1.4 (van Aubel's[5] Theorem). *In triangle ABC let D, E, F be points on the sides BC, CA, AB, respectively, such that AD, BE, and CF are concurrent at P. Then*

$$\frac{AP}{PD} = \frac{AE}{EC} + \frac{AF}{FB}.$$

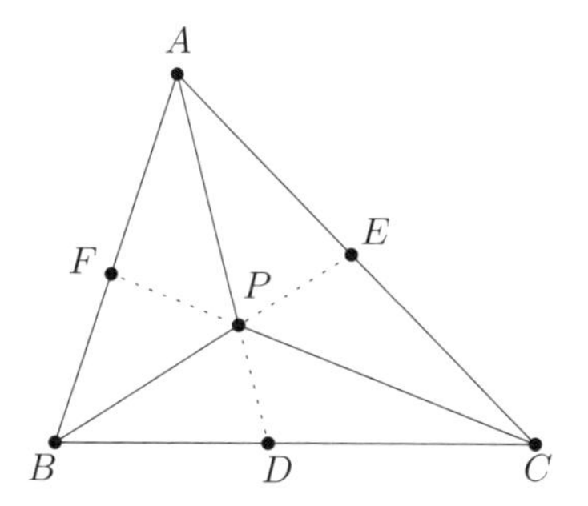

Proof. By the Area Lemma (Proposition 1.27) the right-hand side equals

$$\frac{[APB]}{[BPC]} + \frac{[APC]}{[BPC]} = \frac{[ABPC]}{[BPC]} = \frac{[ABC]}{[BCP]} - 1.$$

It remains to apply the Area Lemma again, this time on the left-hand side:

$$\frac{AP}{PD} = \frac{AD}{PD} - 1 = \frac{[ABC]}{[BCP]} - 1.$$

This concludes the proof. □

Note that if we choose E and F to be the midpoints of the sides AC and AB, we obtain another proof that medians divide each other in the ratio $2:1$ (see Proposition 1.5). Another notable corollary follows if we take P to be the incenter of triangle ABC.

[5]Henri Hubert van Aubel (1830–1906) was a Dutch professor of mathematics.

Corollary 1.28. *Let ABC be a triangle with incenter I and let $AI \cap BC = D$. Then:*

$$\frac{AI}{ID} = \frac{b+c}{a} \quad \text{and} \quad \frac{ID}{AD} = \frac{a}{a+b+c}.$$

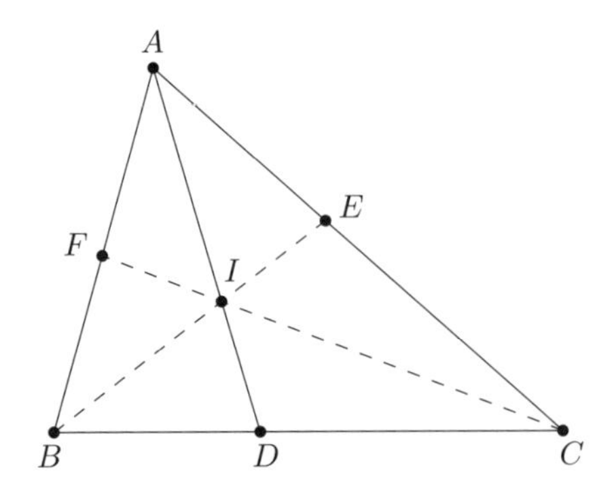

Proof. Let $E = BI \cap AC$ and $F = CI \cap AB$. Then van Aubel's Theorem combined with the Angle Bisector Theorem imply

$$\frac{AI}{ID} = \frac{AE}{EC} + \frac{AF}{FB} = \frac{c}{a} + \frac{b}{a} = \frac{b+c}{a}.$$

The second relation follows from $ID/AD = 1 + ID/AI$.

□

Example 1.5 (AIME 1989). *Let ABC be a triangle and let D, E, F lie on the sides BC, CA, AB, respectively, such that AD, BE and CF are concurrent at P. Given that $AP = 6$, $BP = 9$, $PD = 6$, $PE = 3$, and $CF = 20$, find the area of triangle ABC.*

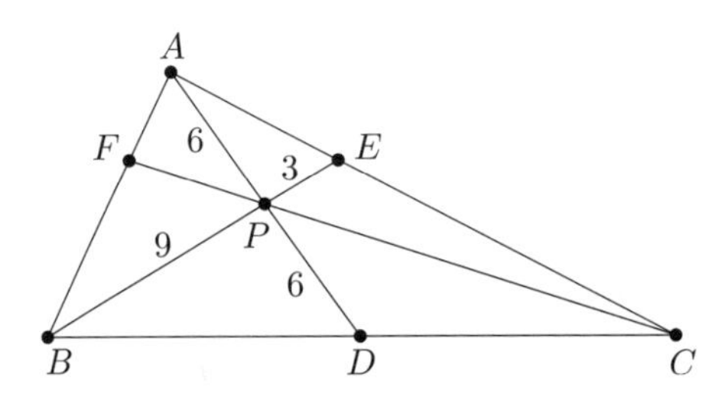

Proof. By the Area Lemma (see Proposition 1.27) we have

$$\frac{[BPC]}{[ABC]} = \frac{DP}{DA} = \frac{1}{2}, \quad \frac{[CPA]}{[ABC]} = \frac{EP}{EB} = \frac{1}{4}.$$

Since $[ABC] = [BPC] + [CPA] + [APB]$, this implies $[APB] = \frac{1}{4}[ABC]$, and using the Area Lemma again we obtain $FP = \frac{1}{4}CF = 5$ and $CP = 15$. Also,

$[ABP] = [CPA]$ implies (Area Lemma yet again!) that D is the midpoint of BC.

Now the median formula from Corollary 1.24 in triangle BCP yields

$$PD^2 = \frac{CP^2 + BP^2}{2} - \frac{BC^2}{4}.$$

Hence $BC = 6\sqrt{13}$ and applying Heron's formula in triangle BCP with sides 15, 9, and $6\sqrt{13}$ we obtain $[BPC] = 54$ and $[ABC] = 108$. □

Circles, Angles

As we will see in this section, angles are intimately related to many common geometric configurations. Being able to manipulate them smoothly is therefore a skill one has to learn.

In the following proof we will reach the result by computing various angles in the picture. This technique is called *angle-chasing* and it is probably the most frequently used method in Euclidean geometry.

Example 1.6. *Let $ABCD$ be a quadrilateral such that $AB = AC$, $AD = CD$, $\angle BAC = 20°$, and $\angle ADC = 100°$. Show that $AB = BC + CD$.*

Proof. Observe that since triangles BCA and ACD are isosceles, we have

$$\angle CBA = \angle ACB = 80°, \qquad \angle CAD = \angle DCA = 40°.$$

Let P be a point on segment AB such that $PA = AD = CD$. We aim to prove $BP = BC$.

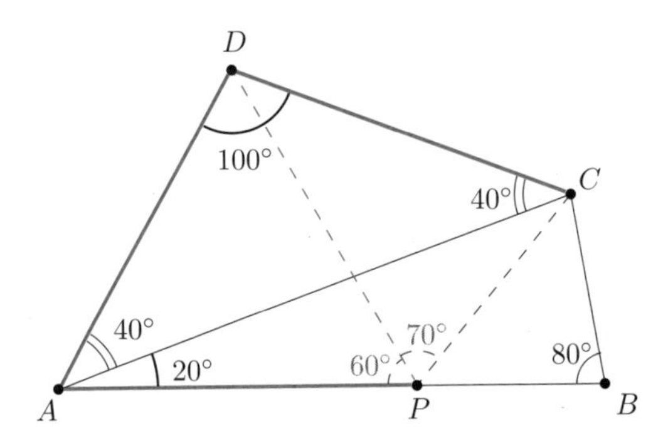

Note that $\angle PAD = 20°+40° = 60°$. Triangle PAD has two equal sides AP and AD that subtend angle $60°$, so it is equilateral. From this we deduce $PD = AD = CD$, so triangle PDC is isosceles and, since $\angle PDC = 100° - 60° = 40°$, we get $\angle DCP = \angle CPD = 70°$.

Finally, turning our attention to triangle BCP we observe that

$$\angle BPC = 180° - 60° - 70° = 50°$$

which together with $\angle CBP = 80°$ implies $\angle PCB = 50°$, so $PB = BC$ indeed. □

Circles

Angle-chasing would not be as important and powerful if we were not able to employ circles. Fortunately, this is possible thanks to the Inscribed Angle Theorem (Theorem 1.2). Recall that a quadrilateral is called *cyclic* (or *inscribed*) if it can be inscribed in a circle.

Proposition 1.29 (The key properties of cyclic quadrilaterals). *Let $ABCD$ be a convex quadrilateral. Then:*

(a) If $ABCD$ is cyclic then any of its sides is visible from the other two vertices under the same angle, and any of its diagonals is visible from the other two vertices under angles that sum up to $180°$.

(b) If there is a side of $ABCD$ that is visible from the other two vertices under the same angle, then $ABCD$ is cyclic.

(c) If there is a diagonal of $ABCD$ that is visible from the other two vertices under the angles that sum up to $180°$, then $ABCD$ is cyclic.

Proof. For (a), denote by O the circumcenter of $ABCD$. By the Inscribed Angle Theorem $\angle ACB = \frac{1}{2}\angle AOB = \angle ADB$, and the other three equalities are proved similarly. Furthermore, as angles by O determined by segments OB and OD sum up to $360°$, angles by A and C add up to $\frac{1}{2}360° = 180°$. Analogously, we obtain $\angle B + \angle D = 180°$.

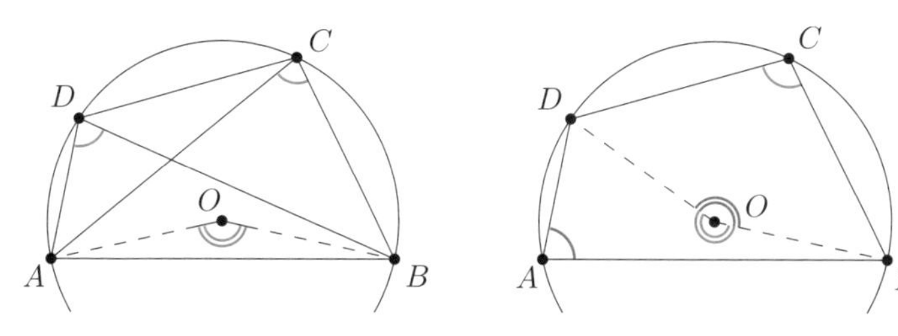

In (b), let us without loss of generality assume $\angle ACB = \angle ADB$, and denote by D' the second intersection of line AD and the circumcircle of triangle ABC. By (a), $\angle AD'B = \angle ACB = \angle ADB$ implying that lines DB and $D'B$ are parallel. Thus points D and D' coincide and quadrilateral $ABCD$ is cyclic.

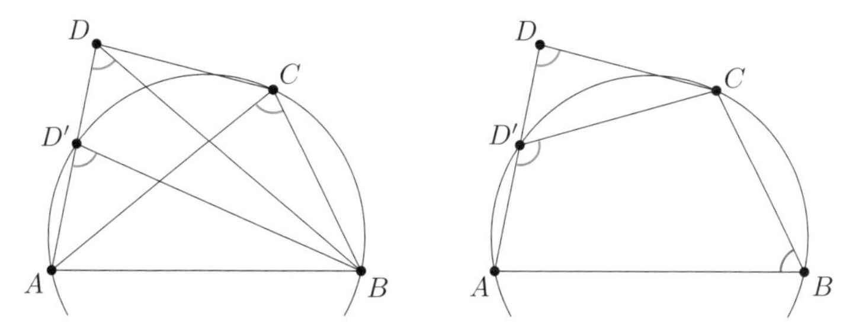

In (c) we proceed similarly. Assume that $\angle ABC + \angle CDA = 180°$ and denote by D' the second intersection of line AD and circumcircle of triangle ABC. Again by (a), $\angle AD'C = 180° - \angle ABC = \angle ADC$, so $CD \parallel CD'$, $D = D'$, and $ABCD$ is cyclic. □

As a direct consequence of the previous proposition we obtain that given a circle ω, its fixed chord AB, and a variable point X on ω, there are only two possible magnitudes of $\angle AXB$ depending on the relative position of line AB and point X. Moreover, these two magnitudes add up to $180°$.

On the other hand, given a segment AB and angle φ, the locus of points X for which $\angle AXB = \varphi$ consists of two circular arcs symmetric with respect to AB.

Also in a configuration with four points on a circle one can find many similar triangles.

Corollary 1.30. *Let $ABCD$ be a cyclic quadrilateral, let P be the intersection of its diagonals and let R be the intersection of rays BA and CD. Then:*

(a) $\triangle ABP \sim \triangle DCP$ and $\triangle BCP \sim \triangle ADP$,
(b) $\triangle RAD \sim \triangle RCB$ and $\triangle RAC \sim \triangle RDB$.

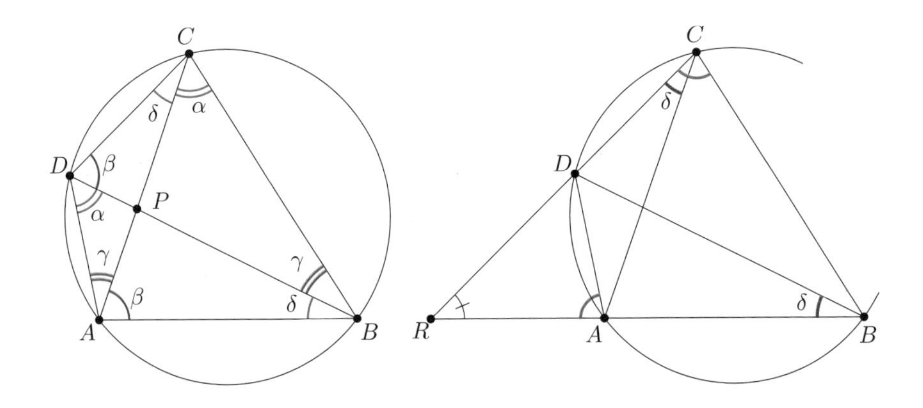

Proof. Divide the circumcircle of $ABCD$ into four arcs AB, BC, CD, DA and denote the corresponding inscribed angles by α, β, γ, δ, respectively. Then $\angle PBA = \delta = \angle DCP$ and $\angle BAP = \beta = \angle PDC$, so triangles ABP and DCP are similar (AA). The second similarity is proved in the same fashion.

In part (b) note that $\angle DAR = 180° - \angle BAD = \angle RCB$. Since triangles RAD and RCB also share an angle, they are similar. Finally, $\angle RCA = \delta = \angle DBR$ which implies $\triangle RAC \sim \triangle RDB$ (AA), and we are done. $\square$

Note that the triangles are similar indirectly. We will discuss this further in the section concerning antiparallelism.

Now, we establish an important corollary of the Inscribed Angle Theorem.

Corollary 1.31 (Correspondence between arcs and angles)**.** *Let AB and CD be equal arcs of a circle ω. Then the inscribed angles corresponding to these arcs are equal.*

Proof. As the arcs are equal, they occupy the same portion of the perimeter of ω and the corresponding central angles are equal. Therefore the corresponding inscribed angles are also equal. □

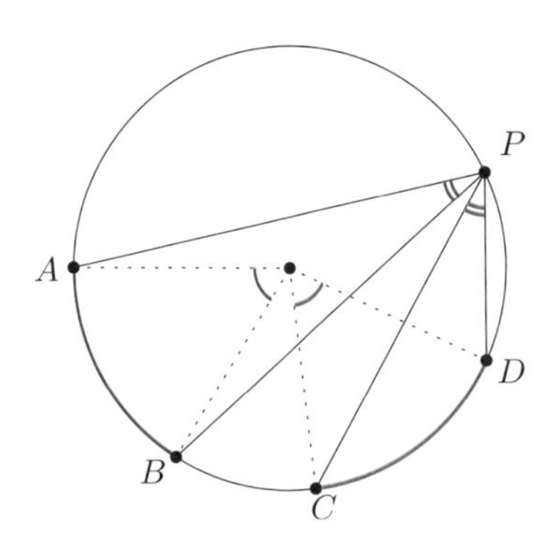

We illustrate this corollary by two examples.

Example 1.7. *Let $ABCD$ be a cyclic quadrilateral.*

(a) If $AD = BC$ then $ABCD$ is a trapezoid.
(b) If $AC = BD$ then $ABCD$ is a trapezoid.

Proof. (a) From $AD = BC$ we infer that the minor arcs AD and BC are equal implying that $\angle DCA$ and $\angle BAC$ are equal. Hence $AB \parallel CD$ and $ABCD$ is a trapezoid.

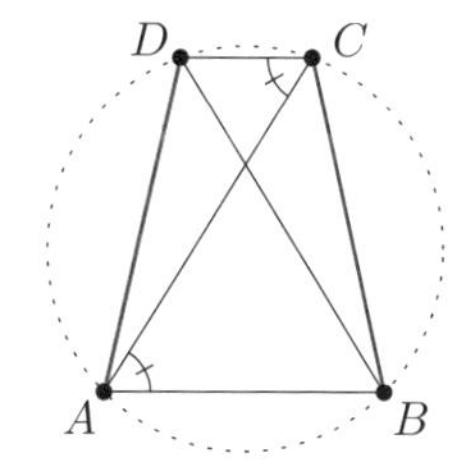

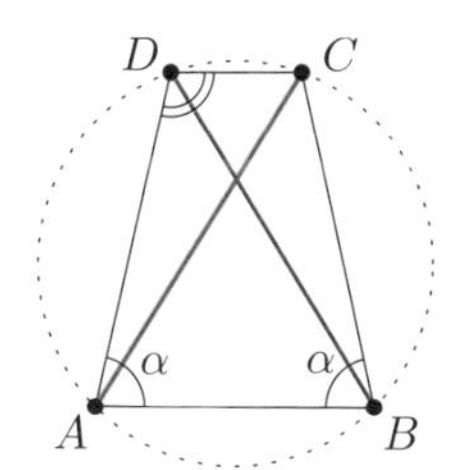

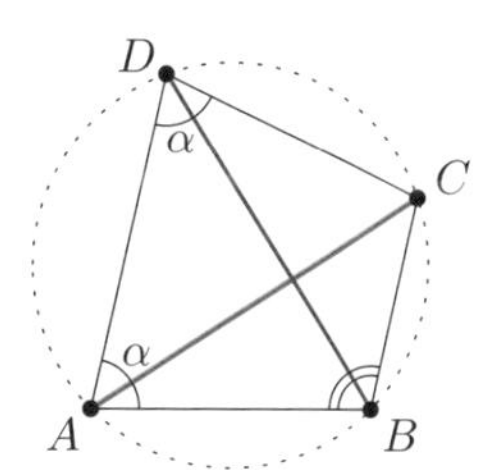

(b) Let $\angle BAD = \alpha$. As $AC = BD$, one of the angles $\angle CBA$, $\angle ADC$ subtends the same arc as the angle $\angle BAD$, and is therefore equal to α too.
If $\angle CBA = \angle BAD = \alpha$ then $\angle ADC = 180° - \angle CBA = 180° - \alpha$ and the lines AB and CD are parallel. Similarly, if $\angle ADC = \angle BAD = \alpha$ then $\angle CBA = 180° - \alpha$ and AD and BC are parallel. □

Example 1.8. *Let ABC be a triangle inscribed in a circle ω and let M_a be the midpoint of arc BC of ω that does not contain point A. Then the internal angle bisector of $\angle A$ and the perpendicular bisector of BC pass through M_a.*

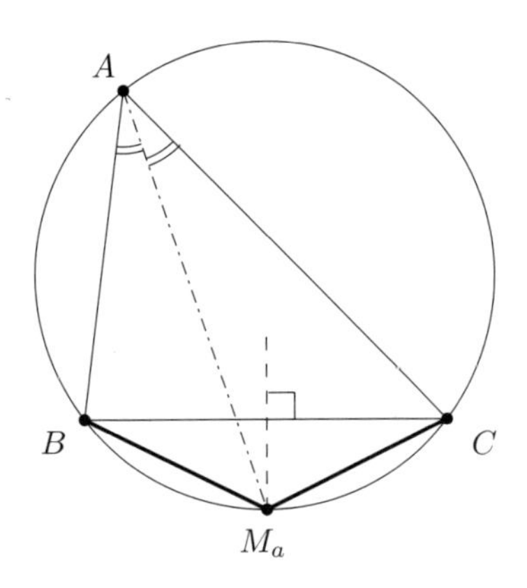

Proof. Since M_a is the midpoint of arc BC, we have $BM_a = CM_a$, so M_a lies on perpendicular bisector of BC. Also, as equal arcs are intercepted by equal angles, we have $\angle BAM_a = \angle M_aAC$. Thus, M_a lies on the angle bisector of $\angle A$. □

The following corollary makes angle-chasing in cyclic quadrilateral easier.

Corollary 1.32. *Let $ABCD$ be a quadrilateral inscribed in a circle ω and denote by P the intersection of its diagonals. Suppose that rays BA and CD intersect at R. Finally, denote the inscribed angles corresponding to arcs BC, DA (not containing A, B) by β, δ. Then*

(a) $\angle BPC = \beta + \delta$,
(b) $\angle BRC = \beta - \delta$.

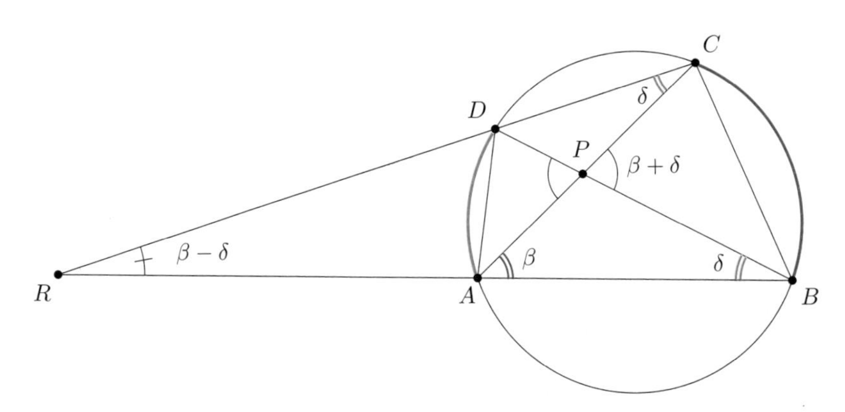

Proof. For part (a) using the property of an exterior angle in triangle ABP we get

$$\angle BPC = \angle BAP + \angle PBA = \beta + \delta.$$

Similarly in part (b), looking at triangle ACR we obtain

$$\angle BRC = \angle BAC - \angle RCA = \beta - \delta.$$

Thus, the corollary is proven. □

This corollary in fact asserts that we can express angle between two chords of a circle in terms of inscribed angles corresponding to some arcs of that circle. With this result we could serve the following proof without words!

Example 1.9. *Let AB be a chord of a circle ω and let M be the midpoint of arc AB. Let line l passing through M intersect the chord AB at P and circle ω for the second time at Q. Similarly, let line m ($m \neq \ell$) passing through M intersect the chord AB and circle ω at R, S, respectively. Show that points P, Q, R, S lie on a circle.*

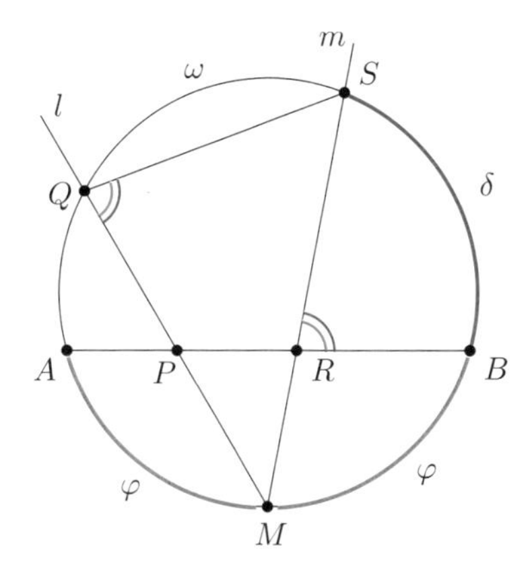

Proof. It is enough to prove that $\angle PQS + \angle SRP = 180^\circ$ or equivalently $\angle MQS = \angle BRS$. Observe that shorter arcs MA and MB of ω are equal, and denote the corresponding inscribed angle by φ. Also denote by δ the inscribed angle corresponding to arc BS of ω not containing A. By Corollary 1.32(a) $\angle BRS = \varphi + \delta = \angle MQS$, so we are done. □

The same result holds even if one of the lines (say l) intersects the line AB outside the circle. We leave the proof as an exercise for the reader, this time with the help of Corollary 1.32(b).

If one wants to prove that three curves pass through a common point, it is often convenient to define the intersection of two of these curves and show that it lies on the third one (a method we have already seen in use in the proofs of the Propositions 1.6 and 1.7).

Theorem 1.33 (Miquel's[6] pivot theorem). *Let P, Q, R be arbitrary points on the sides BC, CA, AB of a triangle ABC. Show that the circumcircles of triangles ARQ, BPR, and CQP pass through a common point.*

[6]Auguste Miquel was a French mathematician active in the mid-nineteenth century.

Proof. Denote by M the second intersection of the circumcircles of triangles BPR and CQP and suppose it lies inside triangle ABC.

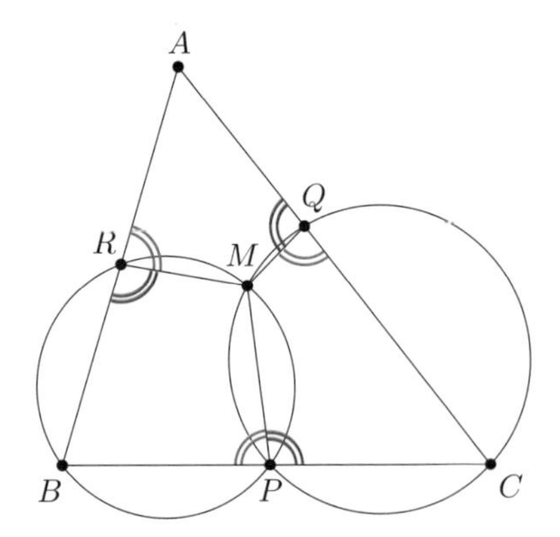

Since quadrilaterals $BPMR$ and $CQMP$ are cyclic, we have

$$\angle MRA = 180^\circ - \angle BRM = \angle MPB = 180^\circ - \angle CPM = \angle MQC = \\ = 180^\circ - \angle AQM.$$

Hence in quadrilateral $ARMQ$ one pair of opposite angles adds up to 180°, so the quadrilateral is cyclic too and the result follows.

The cases when M does not lie inside triangle ABC are treated similarly. □

As we have seen, cyclic quadrilaterals in some sense "multiply" information about angles. Therefore it is worthwhile to look for them when direct angle-chasing seems hopeless.

Example 1.10 (All-Russian Olympiad 1996). *Points E and F are chosen on the side AB of a convex quadrilateral $ABCD$ such that $AE < AF$. Given that $\angle ADE = \angle FCB$ and $\angle EDF = \angle ECF$, prove that $\angle FDB = \angle ACE$.*

Proof. From $\angle EDF = \angle ECF$ we immediately infer that the quadrilateral $CDEF$ is cyclic.

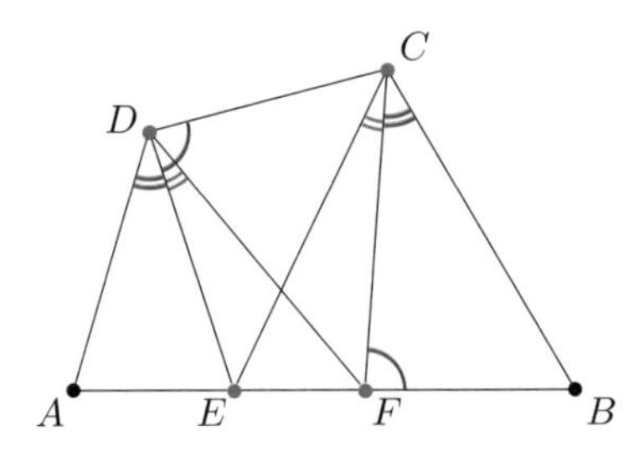

Hence $\angle BFC = \angle EDC$ and

$$180^\circ - \angle CBA = \angle BFC + \angle FCB = \angle EDC + \angle ADE = \angle ADC.$$

The quadrilateral $ABCD$ is then also cyclic. Thus $\angle ADB = \angle ACB$ and the result follows after subtracting $\angle ADF = \angle ECB$. □

Tangents

Another phenomenon that can be characterized by angles is tangency.

The crucial result concerning tangents says that the angle between chord AB of a circle and the tangent at A equals the inscribed angle corresponding to arc AB. We formalize it in the following proposition.

Proposition 1.34. *Let ABC be a triangle inscribed in a circle ω. Let ℓ be a line passing through A different from AB. Let L be a point on ℓ such that AB separates points C, L. Then AL is tangent to ω if and only if $\angle LAB = \angle ACB$.*

Proof. Clearly there is only one line passing through A tangent to ω and there is only one line passing through B that satisfies $\angle LAB = \angle ACB$. Hence it is enough to prove that if AL is tangent to ω, then $\angle LAB = \angle ACB$.

Denote by O the center of ω and by M the midpoint of AB.

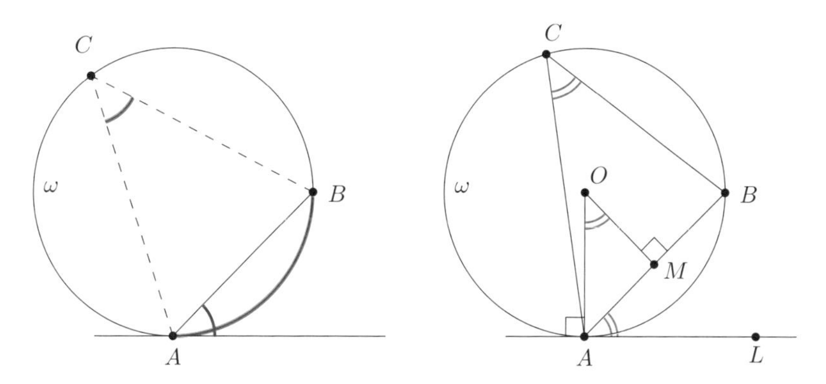

Since AL is tangent to ω, it is perpendicular to the radius OA. This implies

$$\angle LAB = 90^\circ - \angle MAO = \angle AOM = \frac{1}{2}\angle AOB = \angle ACB,$$

and we are done. □

Example 1.11. *Let ABC be a triangle. Denote by ω_a the circle tangent to AB at A and passing through C. Similarly, denote by ω_b the circle tangent to BC at B and passing through A, and ω_c the circle tangent to CA at C and passing through B. Prove that the circles ω_a, ω_b, ω_c intersect at one point.*

Proof. Denote by K the second intersection of circles ω_a and ω_b.

Since BC is tangent to ω_b at B, we get $\angle CBK = \angle BAK$. Similarly, looking at circle ω_a and its tangent AB we obtain $\angle BAK = \angle ACK$. Putting together gives $\angle CBK = \angle ACK$.

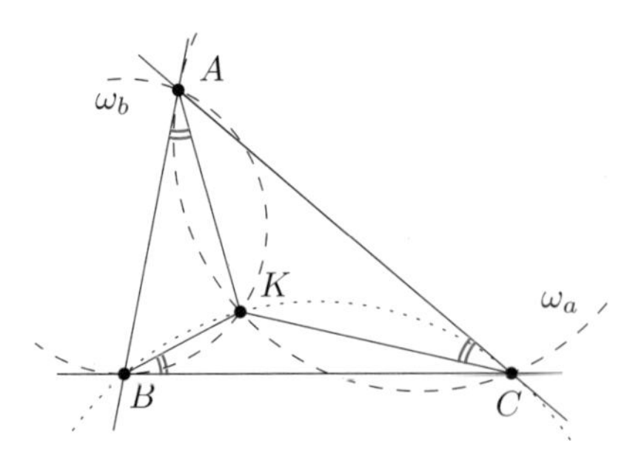

But this implies that CA is tangent to the circumcircle of triangle BCK. Since there is a unique circle passing through B and tangent to AC at C, we get that the circumcircle of BCK *is* in fact ω_c. So K lies on ω_c and the result follows. □

The point K is called the first Brocard[7] point of a triangle ABC. The second Brocard point is obtained by intersecting circles defined by reversed order of letters A, B, C.

If two circles are tangent, drawing their common tangent at that point can often do the trick.

Example 1.12. *Circles ω_1 and ω_2 are tangent externally at point T. Their common external tangent t is tangent to them at A, B, respectively. Show that $\angle ATB = 90°$.*

Proof. Denote by M the intersection of line t and the common internal tangent of circles ω_1, ω_2.

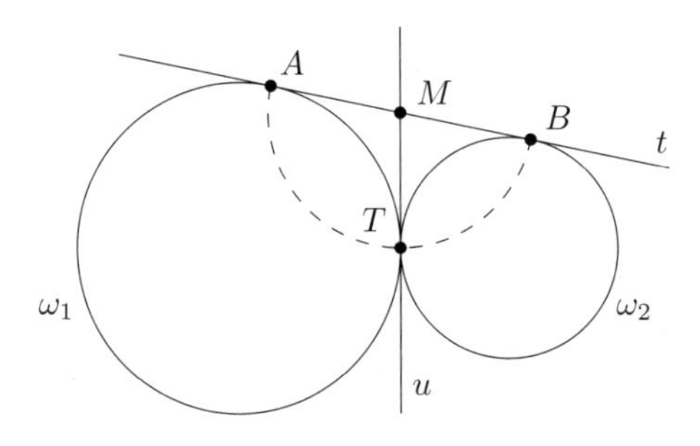

Tangents from M to circles ω_1, ω_2 are equal, so we have $MA = MT = MB$. Thus M is the midpoint of AB and also the circumcenter of triangle ABT implying $\angle BTA = 90°$ as desired. □

[7]Pierre René Jean Baptiste Henri Brocard (1845–1922) was a French mathematician and meteorologist. He is regarded to be one of the co-founders of modern triangle geometry.

Antiparallel lines

In this section we shall discuss another angle-chasing technique. Although this part may seem complicated without bringing immediate reward, the insight it can give is invaluable. We strongly encourage the reader to pay close attention.

When two similar triangles share one angle, the corresponding sides opposite to the common angle may be either parallel, if the similarity is direct, or they may form a cyclic quadrilateral as we have seen in Corollary 1.30, if the similarity is indirect.

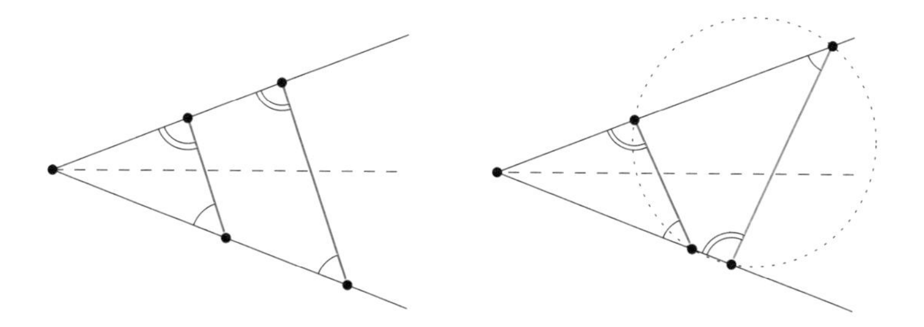

We see that in this case direct and indirect similarity are only one reflection "apart", namely the reflection about the bisector of the common angle. The concept of antiparallel lines makes use of the connection between indirect similarity and cyclic quadrilaterals.

Now we are ready to form a definition. Given a line n we say that lines ℓ and m (neither parallel to n) are antiparallel with respect to line n if the reflection ℓ' of ℓ about n is parallel to m. When the line n is understood we often omit specific reference to it when discussing antiparallelism. Observe that the following holds:

(a) If ℓ is antiparallel to m then it is antiparallel to all lines parallel to m.
(b) (Symmetry) If ℓ is antiparallel to m then m is antiparallel to ℓ.
(c) Given a line n and a set of mutually parallel lines, then lines antiparallel to all of these form again a set of mutually parallel lines.

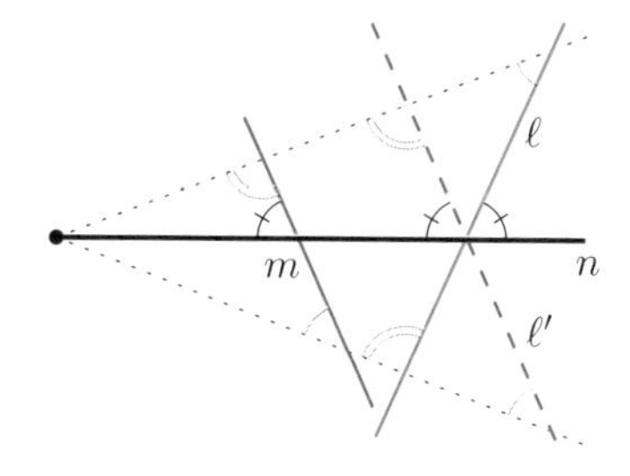

The idea should now be clear, it remains to formalize it and describe all possible cases. As we will see, tangency (as a limit case of concyclicity) will also be involved.

Proposition 1.35. *Let line m intersect ray OA, OB of angle AOB at distinct points X, Y, respectively. Let line ℓ, ($\ell \neq m$) intersect lines OA, OB of angle AOB at (not necessarily distinct) points P, Q, respectively. Then ℓ and m are antiparallel with respect to the angle bisector of angle AOB if and only if one of the following (based on the configuration) holds:*

(a) Points X, Y, P, Q are concyclic (if they are pairwise distinct).
(b) Line OA is tangent to the circumcircle of triangle XYQ (if $X = P$). Similar result holds if $Y = Q$.
(c) Line ℓ is tangent to the circumcircle of triangle XYO (if ℓ passes through O).

Proof. Assume first that ℓ and m are antiparallel. Denote the bisector of $\angle AOB$ by n. If line m is perpendicular to n then ℓ is also perpendicular to n, and the conclusion is clear. Otherwise, denote by m' line symmetric to m with respect to n, and by X', Y' its intersections with OB, OA, respectively.

For part (a), there are four cases to consider corresponding to choices ℓ_1 to ℓ_4, and P_1, Q_1 to P_4, Q_4 (see diagram). In each of them we get $\angle OPQ = \angle OY'X' = \angle OYX$, and the proposition holds.

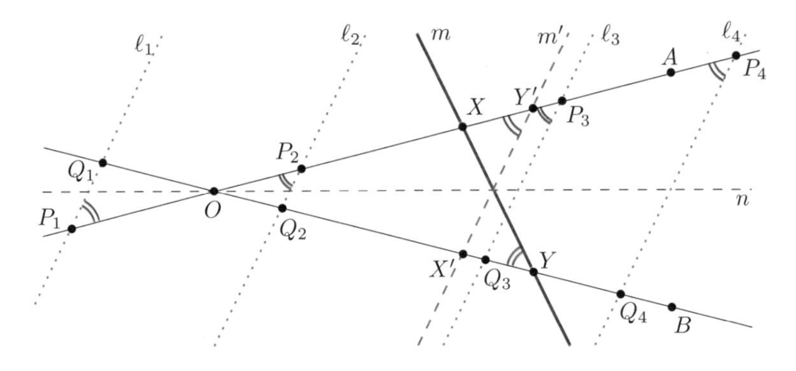

In part (b) we analogously obtain $\angle OP_5Q_5 = \angle OY'X' = \angle OYX$, and using Proposition 1.34 we are done.

Part (c) is proved by Proposition 1.34 as well.

The converse is proved in the same vein. □

Note that if we work with a stripe and its axis instead of angle AOB with its bisector, the previous proposition is valid trivially.

Since antiparallel lines are usually taken with respect to the angle bisector of some angle, let us in that case call these lines *antiparallel with respect to*

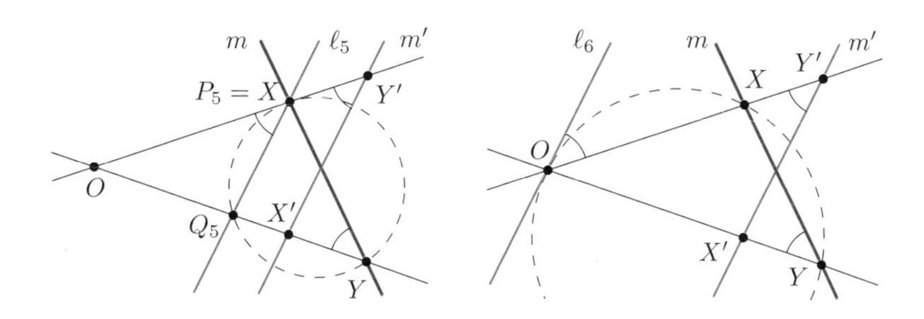

that angle or simply *antiparallel in* that angle. Of particular interest are antiparallel lines that both pass through the vertex of an angle – such lines are called *isogonal*.

Following two examples are both simple and instructive.

Example 1.13. *In a cyclic quadrilateral $ABCD$ let $P = AC \cap BD$, $Q = AD \cap BC$, and $R = AB \cap CD$. Denote by p, q, r the angle bisectors of angles $\angle APB$, $\angle AQB$, $\angle BRC$, respectively. Prove that r is perpendicular to both p and q.*

Proof. We may assume r is horizontal. Since $ABCD$ is cyclic, the lines AC and BD are antiparallel with respect to r, so they form with r an isosceles triangle. In an isosceles triangle with horizontal base, the angle bisector is vertical. Hence $r \perp p$.

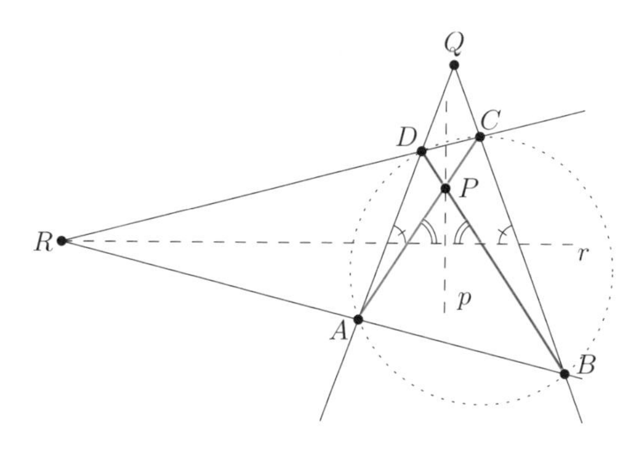

At the same time, $ABCD$ is formed by lines AD and BC. Therefore AD and BC are also antiparallel with respect to r, and similarly as before, $r \perp q$. □

This example has an interesting consequence.

Corollary 1.36. *In a cyclic quadrilateral $ABCD$, let $P = AC \cap BD$, $Q = AD \cap BC$, and $R = AB \cap CD$. Let ℓ, ℓ' be two lines which are antiparallel with respect to one of the angles APB, AQB, BRC. Then they are also antiparallel with respect to the remaining two angles.*

Proof. Clearly, if the lines ℓ, ℓ' are antiparallel with respect to some line o then they are antiparallel with respect to any line o' parallel to o and also with respect to any line o'' perpendicular to o.

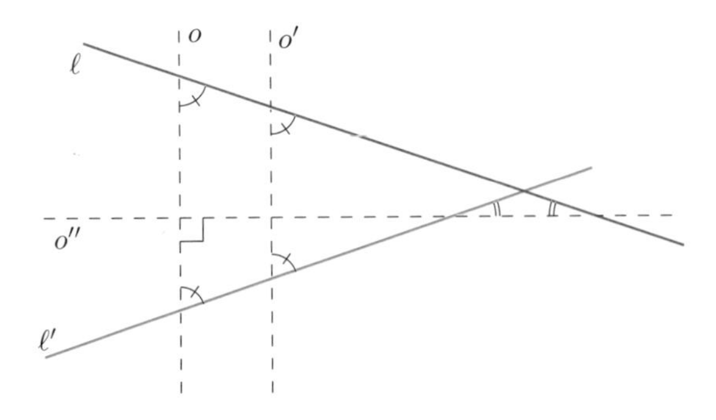

Since any pair of the bisectors of the angles APB, AQB, BRC is either parallel or perpendicular, the conclusion follows. □

Example 1.14 (Czech and Slovak 2010). *Circles ω_1, ω_2 intersect at A, B. Their common external tangent t is tangent to them at K, L, respectively, such that B lies inside triangle KLA. Line ℓ passing through A intersects circles ω_1, ω_2 at M, N, respectively. Prove that ℓ is tangent to the circumcircle of triangle KLA if and only if quadrilateral $KLNM$ is cyclic.*

Proof. Denote by m the angle bisector of lines t and ℓ (or axis of the stripe if they are parallel). Every pair of antiparallel lines will be considered with respect to m. First note that since t is tangent to the circumcircle of triangle KAM, line KA is antiparallel to KM. For similar reasons is LA antiparallel to LN.

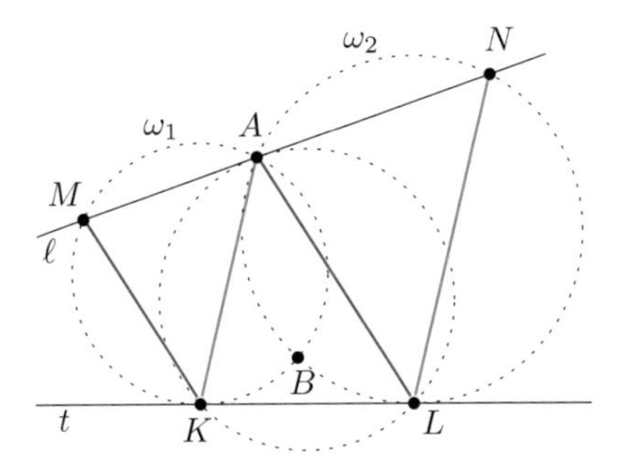

For the if part, if ℓ is tangent to the circumcircle of triangle KLA then KA and AL are antiparallel. Since KM is antiparallel to KA, KA to AL, and AL to LN, together we get that KM is antiparallel to LN. Hence $KLNM$ is cyclic.

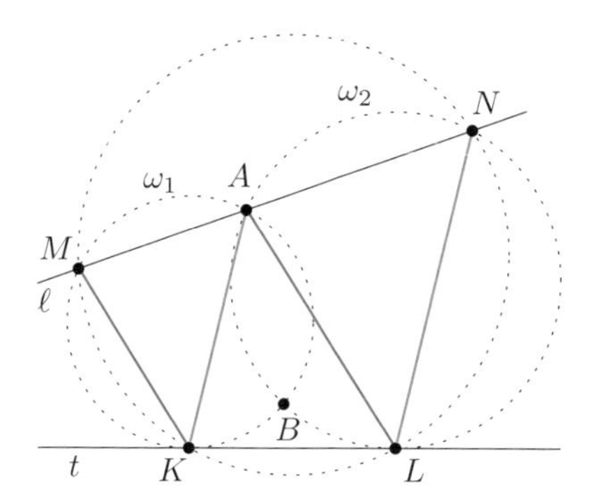

On the other hand, if $KLNM$ is cyclic, AK is antiparallel to KM, KM to LN, and LN to LA. Thus, KA is antiparallel to LA and ℓ is tangent to the circumcircle of triangle KAL. □

Directed angles mod[8] 180°

We end this section by introducing another advanced concept. Some angle-chasing solutions require casework according to the relative position of points (e.g. Is a triangle acute or obtuse? In which half-plane does an intersection lie? In which order do points lie on a line/circle?). This casework can often be shortened if one uses what is called *directed angles mod* 180°.

Magnitude of an angle between lines l, m intersecting at vertex O can be viewed as a number from interval $[0, 180)$ describing (in degrees) the amount of counter-clockwise rotation around O which takes l to m. Let us call this quantity *the directed measure of an angle* and denote it by $\angle(l, m)$. Note that order of lines in brackets matters – in fact $\angle(l, m) + \angle(m, l) = 180°$. If we adopt this point of view, some properties become very neat.

Proposition 1.37. *(a)* $\angle(l, m) + \angle(m, n) = \angle(l, n)$, *with addition mod* 180°.

(b) For any point P $\angle(PA, AB) = \angle(PA, AC)$ *if and only if points* A, B, C *lie on a single line in some order.*

(c) $\angle(AC, CB) = \angle(AD, DB)$ *if and only if points* A, B, C, D *lie on one circle in some order.*

Proof. First two parts are clear, in the first one we just mustn't forget to work mod 180° (see diagram).

The third part is a consequence of Proposition 1.29.

□

Especially, the characterization of cyclic quadrilaterals is very useful. We

[8]This means, we shall work with remainders after division by 180. For example, instead of 200°, we shall work with 20°.

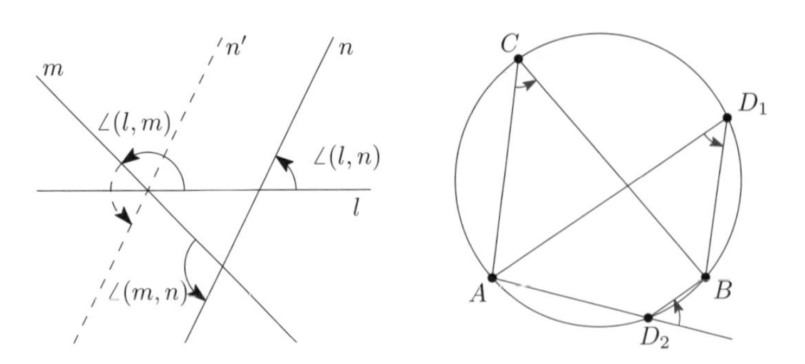

will demonstrate the use of directed angles on one example, where they simplify the casework substantially (check yourself!).

Example 1.15 (Simson[9] line). *Let ABC be a triangle and X a point in its plane. Denote by P, Q, R the feet of perpendiculars dropped from X to the lines BC, CA, AB, respectively. Prove that P, Q, R lie on a single line if and only if X lies on the circumcircle of the triangle ABC.*

Proof. First, assume one of the feet coincides with some vertex of triangle ABC, say P with C.

As Q belongs to AC, the feet P, Q, R are collinear if and only if Q coincides with C or R coincides with A. The first case corresponds to $X = C$, the second to X being antipodal to B, either way the proposition holds. Now let P, Q, R, A, B, C be pairwise distinct.

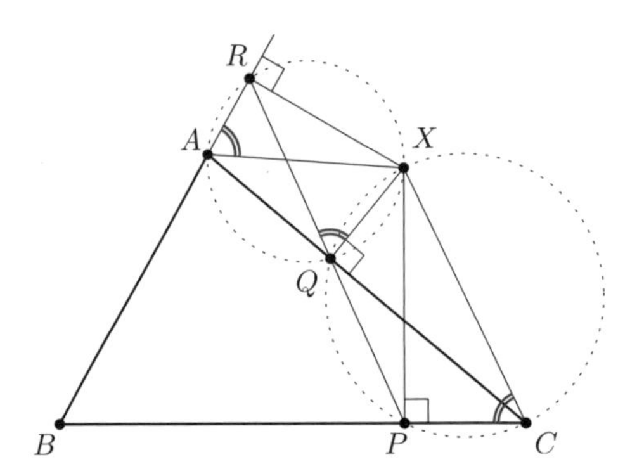

Since P, Q, R are the feet of perpendiculars, we have $\angle(XQ, QA) = 90° = \angle(XR, RA)$. Thus, by Proposition 1.37(c), points X, A, Q, R lie on a circle in some order. Similarly, X, C, Q, P lie on a circle. Hence

$$\angle(XQ, QR) = \angle(XA, AR) = \angle(XA, AB)$$

[9]Robert Simson (1687–1768) was a Scottish mathematician and professor of mathematics at the University of Glasgow.

and

$$\angle(XQ, QP) = \angle(XC, CP) = \angle(XC, CB).$$

Therefore $\angle(XQ, QR) = \angle(XQ, QP)$ holds if and only if $\angle(XA, AB) = \angle(XC, CB)$ does. The former is equivalent to P, Q, R being collinear, the latter to A, B, C, X being concyclic. The result follows. □

For additional practice give new proofs which avoid casework to Theorem 1.2 and to Theorem 1.33 with points P, Q, R on the sidelines of triangle ABC (not neccessarily on the triangle sides).

Ratios

Let's start with two examples!

Example 1.16. *Let P and Q be distinct points on the same side of line ℓ and let X and Y, respectively, be their projections onto ℓ. Denote by Z the intersection of lines YP and XQ. If $PX = 4$ and $QY = 6$, find the distance from Z to line ℓ.*

Proof. Let Z_0 be the projection of Z onto line ℓ. Since $PX \parallel QY$, we see that $\triangle PZX \sim \triangle YZQ$ with factor $k = QY/PX = \frac{3}{2}$. Also, as $ZZ_0 \parallel QY$, we have $\triangle XZZ_0 \sim \triangle XQY$ and we also can find the factor of similarity, since

$$\frac{QX}{ZX} = 1 + \frac{ZQ}{ZX} = 1 + k = \frac{5}{2}.$$

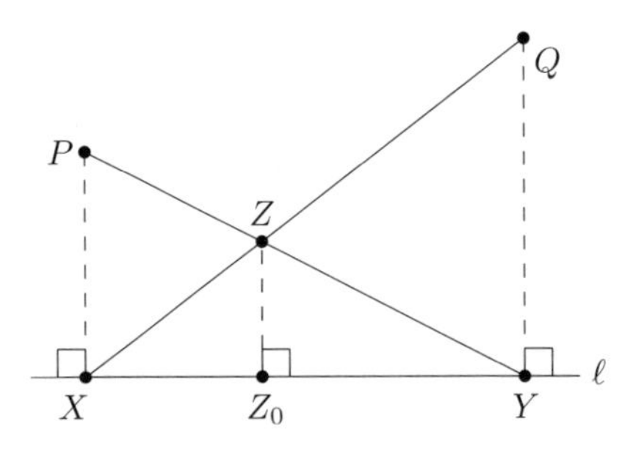

From here, we easily deduce that $ZZ_0 = \frac{2}{5}QY = \frac{12}{5}$. □

Example 1.17. *Let ABC be a right triangle with $\angle A = 90°$ and AD ($D \in BC$) its altitude. Denote by r, r_1, r_2 the inradii of triangles ABC, ABD, ACD, respectively. Prove that*

$$r^2 = r_1^2 + r_2^2.$$

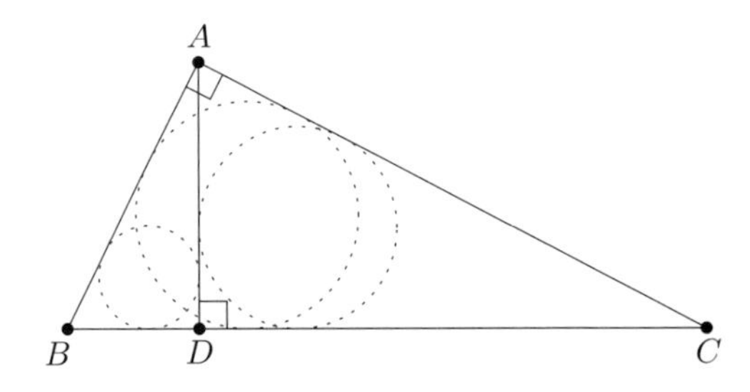

Proof. Triangles ABC, DBA, DAC are similar (AA), meaning that they are also proportional. In particular, the ratio of inradius over hypotenuse is the same number k for all three of them. Therefore we have

$$r = k \cdot BC, \quad r_1 = k \cdot AB, \quad r_2 = k \cdot AC.$$

Hence we are in fact proving

$$k^2 \cdot BC^2 = k^2 \cdot AB^2 + k^2 \cdot AC^2,$$

which is just the Pythagorean Theorem in triangle ABC. □

We have seen how working with ratios via similarities may lead to solution. Now we are going to develop strong connections between ratios and basic geometric concepts such as concyclicity, collinearity and concurrence.

Power of a Point

Proposition 1.38. *(a) Let $ABCD$ be a convex quadrilateral and let $P = AC \cap BD$. Then the points A, B, C, D are concyclic if and only if*

$$PC \cdot PA = PB \cdot PD.$$

(b) Let $ABCD$ be a convex quadrilateral and let $P = AB \cap CD$. Then the points A, B, C, D are concyclic if and only if

$$PA \cdot PB = PC \cdot PD.$$

(c) Assume points P, B, C are collinear in this order and point A does not lie on this line. Then the line PA is tangent to the circumcircle of triangle ABC if and only if

$$PA^2 = PB \cdot PC.$$

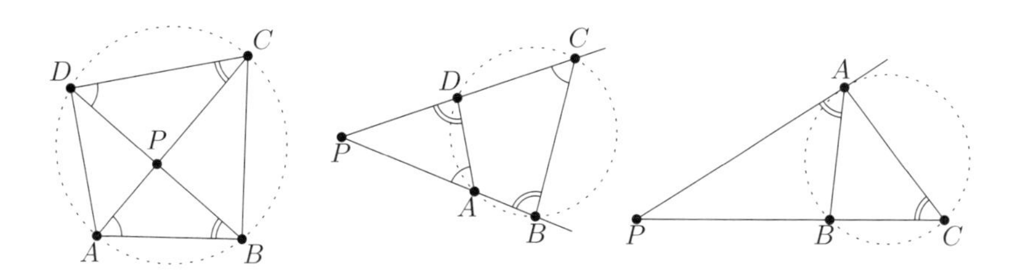

Proof. For part (a), observe that the metric condition can be rewritten as $PA/PB = PD/PC$ and is therefore equivalent to similarity of triangles PAB and PDC (SAS). On the other hand, concyclicity of A, B, C, D is by AA also equivalent to this similarity (see Propositions 1.29 and 1.30(a)), which finishes the proof of (a).

Part (b) is proved in the same fashion using similarity of triangles PAD and PCB.

For (c), line PA is tangent to the circumcircle of triangle ABC if and only if $\angle ACB = \angle PAB$ (recall Proposition 1.34). This is equivalent to $\triangle PAB \sim \triangle PCA$ (AA). Just like in part (a) we deduce that the metric condition is also equivalent to this similarity and we may conclude. □

Thanks to this proposition we can numerically describe the relationship between a circle and a point.

Theorem 1.39 (Power of a Point). *Given point P and circle ω, let ℓ be an arbitrary line passing through P and intersecting ω at points A and B. Then the value of $PA \cdot PB$ does not depend on the choice of ℓ. Also, if P lies outside of ω and PT, $T \in \omega$, is a tangent to ω then $PA \cdot PB = PT^2$.*

If we denote the center of ω by O and its radius by R then $PA \cdot PB = |OP^2 - R^2|$. The quantity

$$p(P, \omega) = OP^2 - R^2$$

is called the power of point P with respect to circle ω.

Proof. The first part is a direct consequence of the previous proposition.

Formula $PA \cdot PB = |OP^2 - R^2|$ follows if we let ℓ pass through O since then we obtain $PA \cdot PB = (OP + R)|OP - R| = |OP^2 - R^2|$. □

Note that the number $p(P, \omega)$ is negative when P lies inside ω, zero when it lies on ω, and positive otherwise.

Now let's reveal how fundamental the concept of the Power of a Point is.

Proposition 1.40. *Let ω_1, ω_2 be two circles with distinct centers O_1, O_2 and radii R_1, R_2, respectively. Then:*

(a) the locus of points X for which $p(X, \omega_1)$ is constant is a circle concentric with ω_1.

(b) the locus of points X for which $p(X, \omega_1) = p(X, \omega_2)$ is a line perpendicular to O_1O_2. This line is called the radical axis of the two circles.

Proof. Part (a) is simple. As $p(X, \omega_1) = XO_1^2 - R_1^2$, we only need XO_1 to be constant. So the locus is indeed a concentric circle (possibly degenerate).

For part (b), we can rewrite the condition as $XO_1^2 - XO_2^2 = R_1^2 - R_2^2$. By the perpendicularity criterion (see Proposition 1.22) such points form a line perpendicular to O_1O_2. □

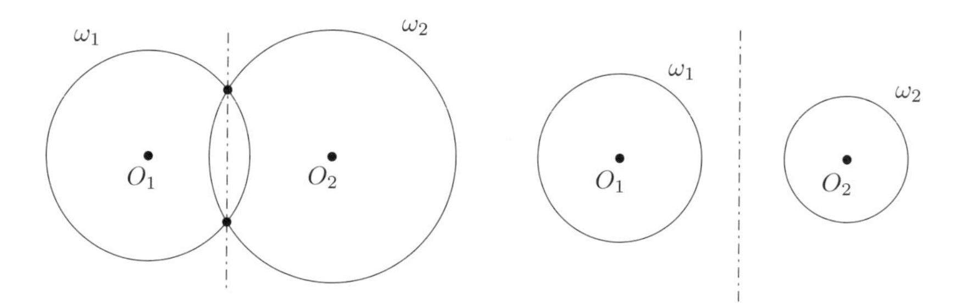

The radical axis is a powerful tool in many problems involving intersecting circles since in that case the radical axis is the line joining their intersections, which both have equal (namely zero) power with respect to the two circles.

Proposition 1.41. *Let the circles ω_1, ω_2 intersect at points A, B. Denote by K, L the points of tangency of the common external tangent with circles ω_1, ω_2, respectively. Then the line AB bisects the segment KL.*

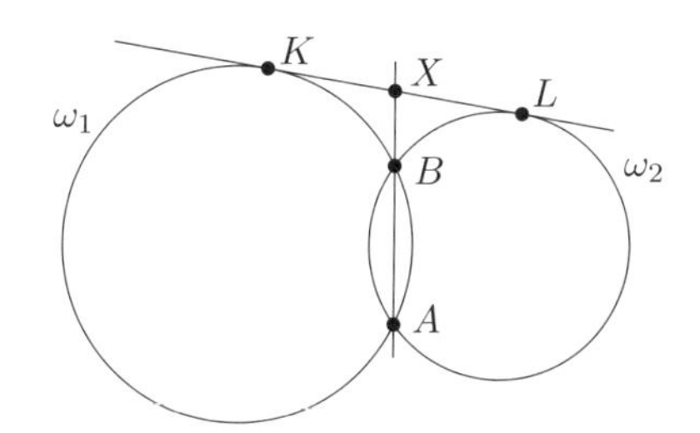

Proof. Let X be the intersection of AB and KL. We know that AB is the radical axis of ω_1 and ω_2, so we may just write

$$XK^2 = p(X, \omega_1) = p(X, \omega_2) = XL^2,$$

and X is the midpoint of KL. □

Proposition 1.42 (Radical center). *Let $\omega_1, \omega_2, \omega_3$ be circles with pairwise distinct centers. Then their pairwise radical axes are either parallel or concurrent. The point of concurrence is called the radical center of the three circles.*

Proof. Assume the radical axis ℓ_1 of circles ω_2, ω_3 intersects the radical axis ℓ_2 of ω_3, ω_1 at point X. Then

$$p(X, \omega_2) = p(X, \omega_3) = p(X, \omega_1),$$

so X lies on the radical axis ℓ_3 of ω_1, ω_2. □

Next we will see a tricky application of this proposition.

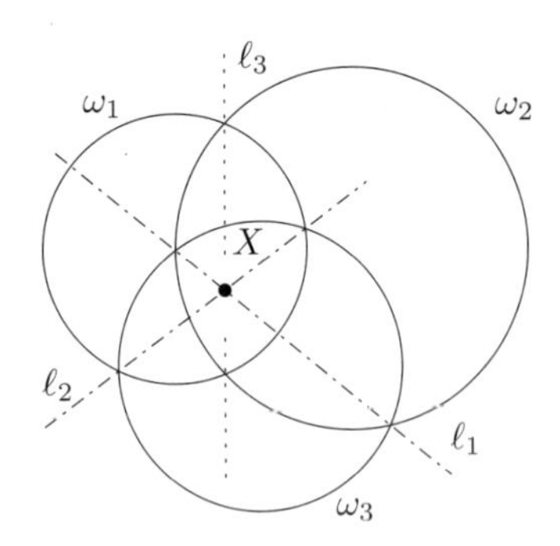

Example 1.18. *Let the incircle ω of triangle ABC touch BC, CA, and AB at D, E, and F, respectively. Let Y_1, Y_2, Z_1, Z_2, and M be the midpoints of BF, BD, CE, CD, and BC, respectively. Let $Y_1Y_2 \cap Z_1Z_2 = X$. Prove that $MX \perp BC$.*

Proof. Consider points B and C as circles with zero radii.

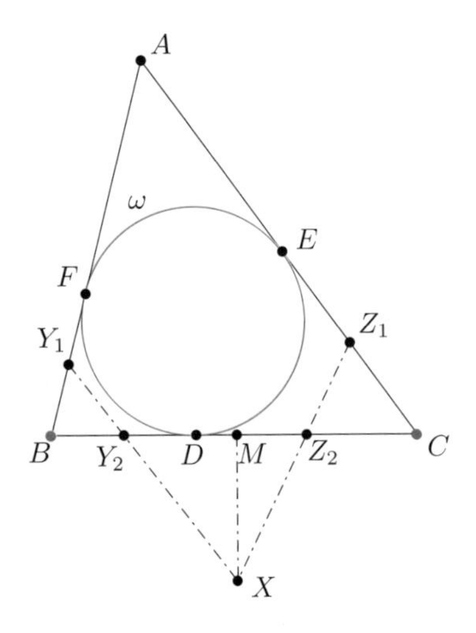

Then

$$p(Y_1, \omega) = Y_1F^2 = Y_1B^2 = p(Y_1, B)$$

and Y_1 lies on the radical axis of ω and B. Similarly, Y_2 lies on this radical axis. Hence the radical axis of ω and B is precisely line Y_1Y_2.

Applying the same argument yields that Z_1Z_2 is the radical axis of ω and C. Thus, X is the radical center of the three circles and lies on the radical axis of B and C, i.e. the perpendicular bisector of BC. □

Now we introduce a more applicable form of the previous theorem, which is the heart of many olympiad problems, since it allows us to translate concurrence to concyclicity and vice versa. Further on, we will refer to it as the *Radical Lemma.*

Proposition 1.43 (Radical Lemma). *Let line ℓ be radical axis of the circles ω_1, ω_2. Let A, D be distinct points on ω_1 and let B, C be distinct points on ω_2 such that the lines AD and BC are not parallel. Then the lines AD and BC intersect at ℓ if and only if $ABCD$ is cyclic.*

Proof. If $ABCD$ is not convex than neither of the conditions can be satisfied and the statement holds. Otherwise, let X be the intersection of AD and BC.

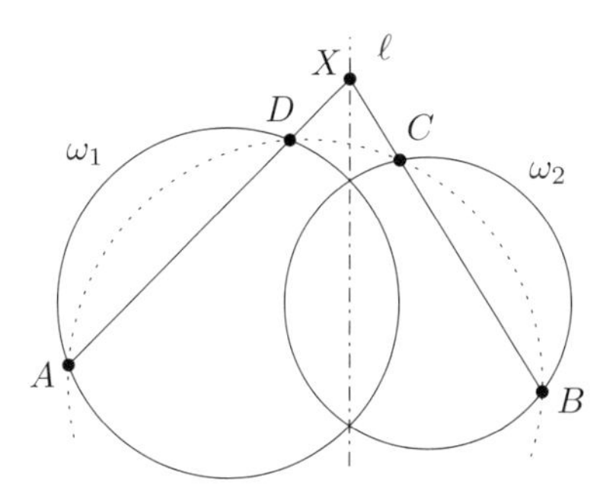

Note that X lies on the radical axis if and only if

$$p(X, \omega_1) = p(X, \omega_2), \quad \text{or equivalently} \quad XD \cdot XA = XC \cdot XB.$$

But the last condition is equivalent to the concyclicity of A, B, C, D (see Proposition 1.38). □

Example 1.19 (IMO 1995). *Let A, B, C, D be four distinct points on a line, in that order. The circles with diameters AC and BD intersect at X and Y. Let P be a point on the line XY such that $P \notin BC$. The line CP intersects the circle with diameter AC at C and M, and the line BP intersects the circle with diameter BD at B and N. Prove that the lines AM, DN, XY are concurrent.*

Proof. First assume that P lies outside the circles. Line XY is the radical axis of the two circles, so the concurrence of BN, CM, and XY by the Radical Lemma implies that points B, C, M, N are concyclic. Using the Radical Lemma in the second direction, we realize it suffices to prove that points A, D, M, N are also concyclic.

But this is easy! We just recall that AC and BD are diameters, write

$$\angle DNM + 90^\circ = \angle BNM = 180^\circ - \angle MCB = \angle DAM + 90^\circ,$$

and conclude that $ADMN$ is cyclic.

The case when P belongs to segment XY is treated similarly. Another option is to regard angles as directed mod 180°. □

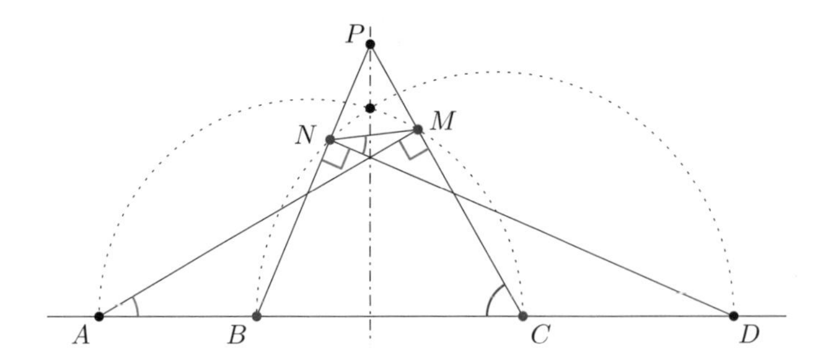

Ceva's[10] Theorem

In this section we shall explore concurrence of the so-called cevians, the segments joining a vertex of a triangle with a point on the opposite side.

Theorem 1.44 (Ceva's Theorem). *Let ABC be a triangle, and let P, Q, R be points on the sides BC, CA, AB, respectively. Then the following assertions are equivalent:*

(a) Lines AP, BQ, CR are concurrent.

(b)

$$\frac{BP}{PC} \cdot \frac{CQ}{QA} \cdot \frac{AR}{RB} = 1 \qquad (\star)$$

(c) (trigonometric form)

$$\frac{\sin\angle PAC}{\sin\angle BAP} \cdot \frac{\sin\angle QBA}{\sin\angle CBQ} \cdot \frac{\sin\angle RCB}{\sin\angle ACR} = 1$$

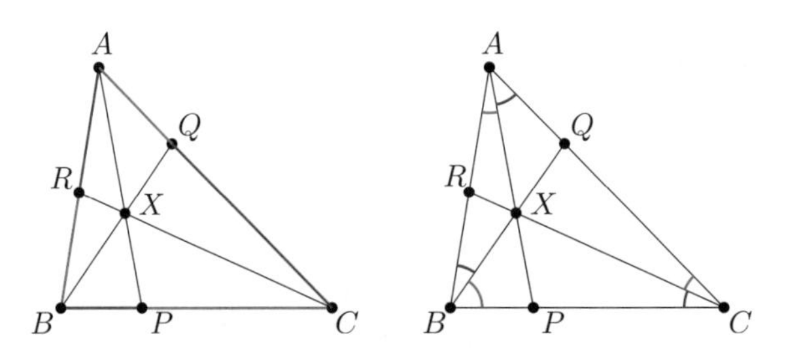

Proof. (a) $\Rightarrow$ (b): Assume the cevians are concurrent at X. Now use Area Lemma (see Proposition 1.27) to learn that

$$\frac{[AXB]}{[AXC]} = \frac{BP}{PC}.$$

[10] Giovanni Ceva (1647–1734) was an Italian mathematician.

Analogously, we obtain

$$\frac{[BXC]}{[AXB]} = \frac{CQ}{QA} \quad \text{and} \quad \frac{[AXC]}{[BXC]} = \frac{AR}{RB}.$$

Multiplying the three relations gives $(\star)$.

(b) $\Rightarrow$ (a): Assume $(\star)$ holds. Intersect BQ and CR at X and also AX and BC at P'. Then AP', BQ, and CR are concurrent cevians, thus

$$\frac{BP'}{P'C} \cdot \frac{CQ}{QA} \cdot \frac{AR}{RB} = 1.$$

Comparing this with $(\star)$ gives

$$\frac{BP'}{P'C} = \frac{BP}{PC},$$

so the points P and P' divide the segment BC in the same ratio and hence they coincide, implying that AP, BQ, CR are concurrent.

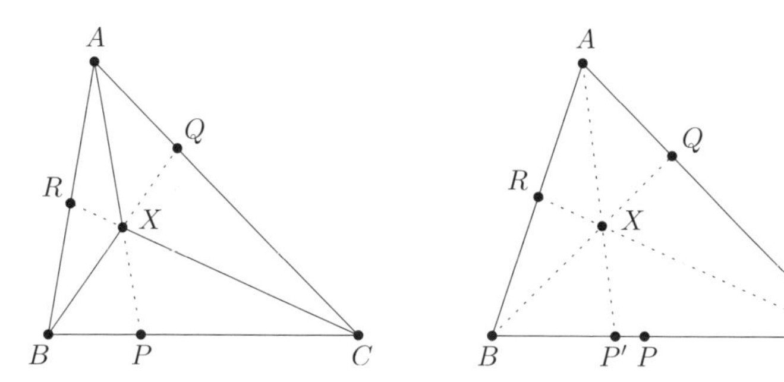

(c) $\Leftrightarrow$ (b): Use the Ratio Lemma (see Proposition 1.18) for adjacent triangles ABP and APC to get

$$\frac{BP}{PC} = \frac{AB \sin \angle BAP}{AC \sin \angle PAC}.$$

Analogously, we obtain

$$\frac{CQ}{QA} = \frac{BC \sin \angle CBQ}{AB \sin \angle QBA} \quad \text{and} \quad \frac{AR}{RB} = \frac{AC \sin \angle ACR}{BC \sin \angle RCB}.$$

After multiplying the three equations, simplifying gives

$$\frac{BP}{PC} \cdot \frac{CQ}{QA} \cdot \frac{AR}{RB} = \frac{\sin \angle BAP}{\sin \angle PAC} \cdot \frac{\sin \angle CBQ}{\sin \angle QBA} \cdot \frac{\sin \angle ACR}{\sin \angle RCB},$$

so the two statements are indeed equivalent.

□

Ceva's Theorem establishes the existence of many important triangle centers, some of which we have already met.

Corollary 1.45. *In triangle ABC the following cevians (always denote their intersections with BC, CA, AB by P, Q, R, respectively) are concurrent:*

(a) medians,
(b) angle bisectors,
(c) altitudes,
(d) cevians corresponding to the points of tangency of the incircle (Gergonne[11] point),
(e) cevians corresponding to the points of tangency of the excircles with the triangle sides (Nagel[12] point),

Proof. (a): As $BP = CP$, $AQ = CQ$, and $AR = BR$, concurrence follows from Ceva's Theorem.

(b): We have $\angle BAP = \angle PAC$, $\angle CBQ = \angle QBA$, and $\angle ACR = \angle RCB$, so the result follows from trigonometric form of Ceva's Theorem.

(c): For altitudes we also use the trigonometric form. We have

$$\frac{\sin \angle BAP}{\sin \angle PAC} = \frac{\sin(90^\circ - \angle B)}{\sin(90^\circ - \angle C)} = \frac{\cos \angle B}{\cos \angle C},$$

and it suffices to multiply three analogous relations to obtain what we need.

(d): In this case $AQ = AR$, $BR = BP$, and $CP = CQ$, so these cevians are concurrent by Ceva's Theorem.

(e): By Proposition 1.15(c) we know that $AR = CP$, $BP = AQ$, and $CQ = BR$, so the concurrence is ensured by Ceva's Theorem again. □

Example 1.20. *Points M, N on the sides AB, AC of the triangle ABC satisfy $MN \parallel BC$. Prove that lines BN and CM intersect on the A-median of triangle ABC.*

Proof. Since $MN \parallel BC$, the sides AB, AC are divided by M, N, respectively, in the same ratio. In other words,

$$\frac{AM}{MB} = \frac{AN}{NC}.$$

Denote by P the intersection of BN and CM and by P' the intersection of AP and BC. Then Ceva's Theorem for concurrent cevians AP', BN, CM implies

$$\frac{BP'}{P'C} \cdot \frac{CN}{NA} \cdot \frac{AM}{MB} = 1, \quad \text{hence} \quad \frac{BP'}{P'C} = \frac{AN}{NC} \cdot \frac{MB}{AM} = 1.$$

[11] Joseph Diaz Gergonne (1771–1859) was a French mathematician and logician.
[12] Christian Heinrich von Nagel (1803–1882) was a German geometer.

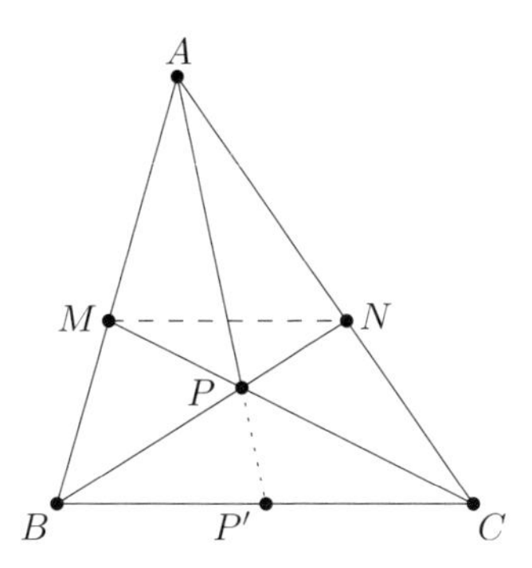

Hence $BP' = P'C$ and P lies on the A-median. □

Theorem 1.46 (Existence of isogonal conjugate). *Let cevians AP, BQ, CR concur at point X. Now construct cevians AP', BQ', CR' which are isogonal to AP, BQ, CR, respectively, in the respective angles. Then the cevians AP', BQ', CR' are concurrent. The point of concurrence is called the isogonal conjugate of X.*

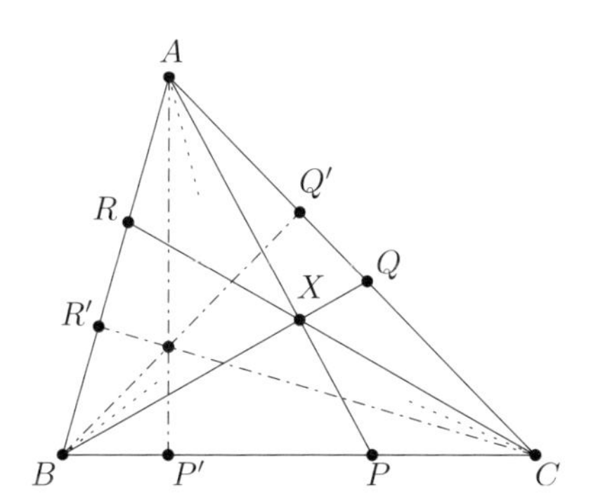

Proof. As AP' is isogonal to AP in $\angle A$, we have $\angle BAP = \angle P'AC$ and $\angle PAC = \angle BAP'$, and similarly for the other cevians. In fact,

$$\frac{\sin\angle BAP}{\sin\angle PAC}\cdot\frac{\sin\angle CBQ}{\sin\angle QBA}\cdot\frac{\sin\angle ACR}{\sin\angle RCB} = \frac{\sin\angle P'AC}{\sin\angle BAP'}\cdot\frac{\sin\angle Q'BA}{\sin\angle CBQ'}\cdot\frac{\sin\angle R'CB}{\sin\angle ACR'}.$$

However, by trigonometric form of Ceva's Theorem, the left-hand side of this equation equals 1 as AP, BQ, and CR are concurrent. Hence also the right-hand side equals 1, and AP', BQ', and CR' are concurrent too. □

We can easily see that the relation of isogonal conjugation is symmetric and except for the incenter, which is the conjugate of itself, it pairs up the points in the triangle. It should be noted that the concept of isogonal conjugation can be easily extended also to points in the exterior of triangle ABC.

We dare to say that one such pair is more important than others. Details are exposed in the next handy little proposition, which will be referred to with a familiarizing name *H and O are friends.*

Proposition 1.47 (*H* and *O* are friends). *Let ABC be a triangle with orthocenter H and circumcenter O. Then the lines AO and AH are isogonal in $\angle A$, and similar result holds for pairs of lines BH, BO and CH, CO, in the respective angles. Therefore H and O are isogonal conjugates.*

Proof. If triangle ABC is acute, we have simply

$$\angle BAO = \frac{1}{2}(180^\circ - \angle AOB) = 90^\circ - \angle C = \angle CAH,$$

and the conclusion follows. In the other cases we proceed similarly. □

Example 1.21 (China MO training 1988). *Let $ABCDEF$ be a hexagon inscribed in a circle ω. Show that the diagonals AD, BE, CF are concurrent if and only if*

$$AB \cdot CD \cdot EF = BC \cdot DE \cdot FA.$$

Proof. Consider the diagonals AD, BE, CF as cevians in triangle ACE, and apply trigonometric form of Ceva's Theorem.

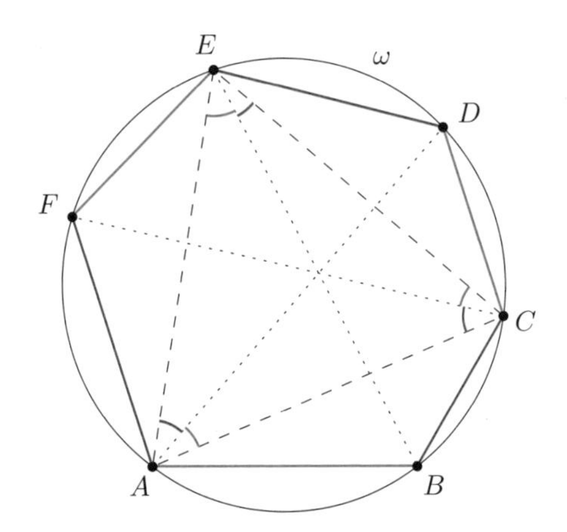

The diagonals are concurrent if and only if

$$\frac{\sin\angle CAD}{\sin\angle DAE} \cdot \frac{\sin\angle ECF}{\sin\angle FCA} \cdot \frac{\sin\angle AEB}{\sin\angle BEC} = 1. \qquad (\clubsuit)$$

Now denote by R the radius of ω. The Extended Law of Sines yields

$$\frac{\sin\angle CAD}{\sin\angle DAE} = \frac{\frac{CD}{2R}}{\frac{DE}{2R}} = \frac{CD}{DE}.$$

In a similar fashion, we obtain

$$\frac{\sin \angle ECF}{\sin \angle FCA} = \frac{EF}{AF} \quad \text{and} \quad \frac{\sin \angle AEB}{\sin \angle BEC} = \frac{AB}{BC}.$$

Plugging this into (♣) and expanding implies the result. □

Menelaus'[13] Theorem

Surprisingly, the criterion for collinearity of three points on the triangle sides (possibly extended) has similar form.

Theorem 1.48 (Menelaus' Theorem). *Let ABC be a triangle and let points D, E, F lie on the lines BC, CA, AB, respectively, so that either none or two of them lie on the triangle sides. Then the points D, E, F are collinear if and only if*

$$\frac{BD}{DC} \cdot \frac{CE}{EA} \cdot \frac{AF}{FB} = 1. \qquad (♠)$$

Proof. Assume first that D, E, F are collinear on a line ℓ and denote by x, y, z the distances of points A, B, C, respectively, from line ℓ. Now using similar triangles yields

$$\frac{y}{z} = \frac{BD}{DC}, \quad \frac{z}{x} = \frac{CE}{EA}, \quad \frac{x}{y} = \frac{AF}{FB}.$$

Multiplying these we obtain (♠).

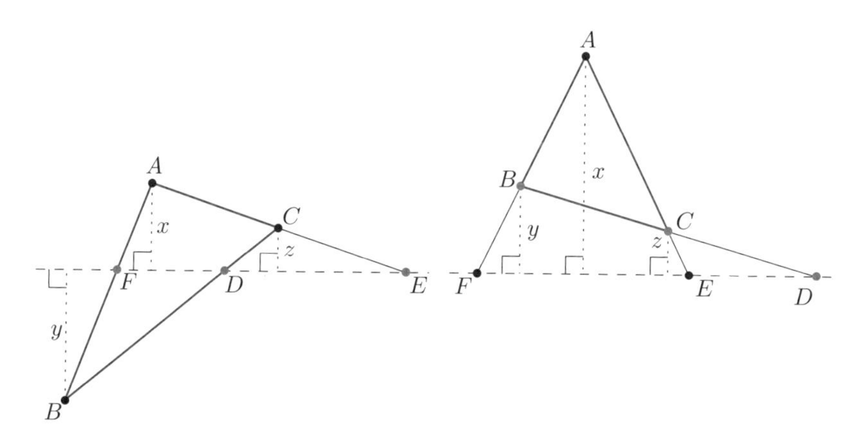

Now assume (♠) holds and let D' be the intersection of EF and BC. Points D', E, F are collinear and we may apply the first part of the statement to obtain

$$\frac{BD'}{D'C} \cdot \frac{CE}{EA} \cdot \frac{AF}{FB} = 1.$$

[13] Menelaus of Alexandria (c. 70–140) was a Greek mathematician and astronomer.

Comparing this with (♠) gives

$$\frac{BD'}{D'C} = \frac{BD}{DC}. \qquad (\diamondsuit)$$

Now realize that both D and D' lie either inside segment BC or outside of it. Either way, ($\diamondsuit$) implies that D and D' coincide, so D, E, F are collinear. □

Thanks to Menelaus' Theorem we can sometimes focus only on a small part of a complicated picture.

Example 1.22. *Let ω be the circumcircle of triangle ABC and let the tangent to ω at A intersect BC at A_1. Define points B_1, C_1 analogously. Prove that A_1, B_1, C_1 are collinear.*

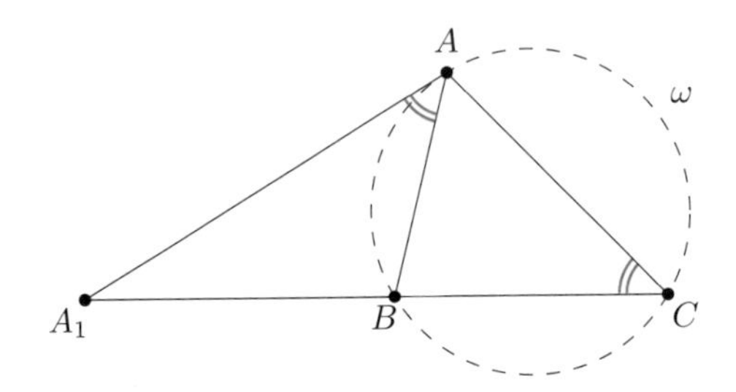

Proof. In order to use Menelaus' Theorem we first calculate A_1B/A_1C from the Ratio Lemma (see Proposition 1.18) in triangle ABC as

$$\frac{A_1B}{A_1C} = \frac{AB}{AC} \cdot \frac{\sin\angle A_1AB}{\sin\angle A_1AC}.$$

Since AA_1 is a tangent, we have $\angle A_1AB = \angle C$ (recall Proposition 1.34) and $\angle A_1AC = 180° - \angle B$. Hence

$$\frac{A_1B}{A_1C} = \frac{AB}{AC} \cdot \frac{\sin\angle C}{\sin\angle B} = \frac{AB^2}{AC^2},$$

having used the Law of Sines in triangle ABC in the last equality. In a similar vein we find

$$\frac{B_1C}{B_1A} = \frac{BC^2}{BA^2} \quad \text{and} \quad \frac{C_1A}{C_1B} = \frac{CA^2}{CB^2}.$$

Multiplying the three fractions gives

$$\frac{A_1B}{A_1C} \cdot \frac{B_1C}{B_1A} \cdot \frac{C_1A}{C_1B} = 1,$$

and Menelaus' Theorem implies the result. □

Example 1.23. *Let ABC be a scalene triangle and M the midpoint of BC. The incircle centered at I touches BC at D. Denote by N the midpoint of AD. Prove that N, I, M are collinear.*

Proof. We may assume $b > c$. Let AA_1 be the internal angle bisector with $A_1 \in BC$. We are going to apply Menelaus' Theorem in triangle ADA_1.

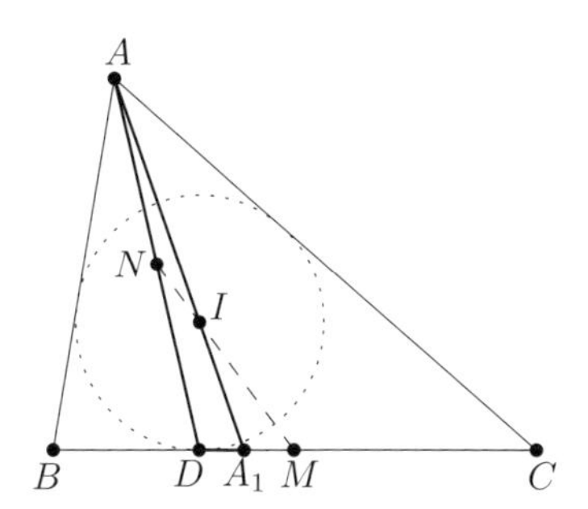

We know that $DN = AN$ and by Corollary 1.28

$$\frac{A_1I}{IA} = \frac{a}{b+c},$$

so it remains to calculate MD and MA_1.

Since $2BD = a + c - b$ (see Proposition 1.15(a)) and M is the midpoint of BC, we get

$$DM = BM - BD = \frac{a}{2} - \frac{a+c-b}{2} = \frac{b-c}{2}.$$

Now using the Angle Bisector Theorem we obtain

$$BA_1 = \frac{ac}{b+c},$$

therefore

$$MA_1 = BM - BA_1 = \frac{a}{2} - \frac{ac}{b+c} = \frac{a(b-c)}{2(b+c)}.$$

Finally, we are ready to use Menelaus' Theorem in triangle ADA_1. Since

$$\frac{AN}{ND} \cdot \frac{DM}{MA_1} \cdot \frac{A_1I}{IA} = 1 \cdot \frac{\frac{b-c}{2}}{\frac{a(b-c)}{2(b+c)}} \cdot \frac{a}{b+c} = 1,$$

the collinearity of N, I, and M follows. □

The following rather subtle example summarizes all techniques discussed in this section.

Example 1.24 (IMO 1995 shortlist). *The incircle of triangle ABC touches the sides BC, CA, AB at points D, E, F, respectively. Let X be a point inside the triangle ABC such that the incircle of triangle XBC touches BC, CX, XB at D, Y, Z, respectively. Show that E, F, Z, and Y are concyclic.*

Proof. First, if $AB = AC$, then D is the midpoint of BC and triangle XBC is also isosceles. Therefore $ZYEF$ is an isosceles trapezoid, hence circumscriptible. Now assume $AB \neq AC$. Since BC is a common tangent of the

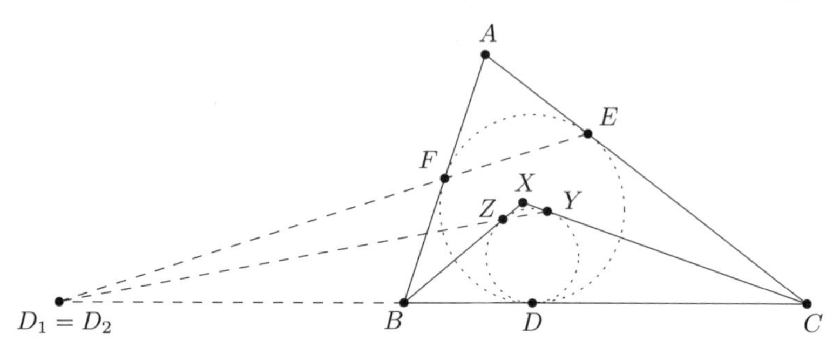

two circles, it is also their radical axis. Hence by the Radical Lemma (see Proposition 1.43) it suffices to prove that BC, EF, and YZ are concurrent. Let $D_1 = EF \cap BC$ and $D_2 = YZ \cap BC$. The key idea is to compare Ceva's Theorem (for triangle ABC and concurrent cevians AD, BE, CF – see Corollary 1.45(d)) and Menelaus' Theorem (for triangle ABC and line EF). We obtain

$$\frac{BD}{DC} \cdot \frac{CE}{EA} \cdot \frac{AF}{FB} = 1 = \frac{BD_1}{D_1C} \cdot \frac{CE}{EA} \cdot \frac{AF}{FB}, \quad \text{hence} \quad \frac{BD}{DC} = \frac{BD_1}{D_1C}.$$

Now we use the same technique for triangle XBC, concurrent cevians XD, BY, CZ, and line YZ. We get

$$\frac{BD}{DC} \cdot \frac{CY}{YX} \cdot \frac{XZ}{ZB} = 1 = \frac{BD_2}{D_2C} \cdot \frac{CY}{YX} \cdot \frac{XZ}{ZB}, \quad \text{hence} \quad \frac{BD}{DC} = \frac{BD_2}{D_2C}.$$

Comparing yields

$$\frac{BD_1}{D_1C} = \frac{BD_2}{D_2C}$$

and since neither of D_1, D_2 lies on segment BC, the points must coincide. Hence BC, EF, and YZ are concurrent and we may conclude. □

Directed segments

Applying Ceva's or Menelaus' Theorem may cause some painful casework as we should be sure about the relative positions of the involved points. However, this may be simplified if we adopt Newton's[14] concept of directed segments.

[14]Isaac Newton (1643–1727) was an English physicist, mathematician and natural philosopher.

A segment emanating from A with endpoint B will be denoted by $\overline{AB}$.

The important property of directed segments is that the ratio or the product of two directed segments, which are part of the same line, is assigned a sign. The sign is positive if the directed segments have the same orientation and negative otherwise. By the same logic we have

$$\overline{AB} = -\overline{BA}.$$

Now we may restate the three important theorems from this chapter in a more general way.

Theorem 1.49 (Power of a Point)**.** *Let a line through P intersect the circle ω at two distinct points A and B. Then*

$$p(P, \omega) = \overline{PA} \cdot \overline{PB}.$$

Theorem 1.50 (Ceva's Theorem)**.** *Let ABC be a triangle and let P, Q, R be points on the lines BC, CA, AB, respectively. Then AP, BQ, CR are concurrent or mutually parallel if and only if*

$$\frac{\overline{BP}}{\overline{PC}} \cdot \frac{\overline{CQ}}{\overline{QA}} \cdot \frac{\overline{AR}}{\overline{RB}} = 1.$$

Theorem 1.51 (Menelaus' Theorem)**.** *Let ABC be a triangle, and let P, Q, R be points on the lines BC, CA, AB, respectively. Then the points P, Q, R are collinear if and only if*

$$\frac{\overline{BP}}{\overline{PC}} \cdot \frac{\overline{CQ}}{\overline{QA}} \cdot \frac{\overline{AR}}{\overline{RB}} = -1.$$

All of them can be proved more or less by copying proofs of their undirected versions. We leave the details to the reader.

Few Notes on Geometric Inequalities

Geometric inequalities form a wide subfield at the border of geometry and algebra. Profound exploration of the area is beyond the scope of this book so we pick and briefly discuss only the inequalities which are most significant or remarkable.

Triangle inequality

There is no doubt that the most important geometric inequality is the renowned triangle inequality.

Theorem 1.52 (Triangle inequality). *Let ABC be a triangle. Then*

$$AB + BC > AC, \qquad BC + CA > BA, \qquad \text{and} \qquad CA + AB > CB.$$

As obvious as triangle inequality may sound, it produces notable results when cleverly applied.

Example 1.25. *Let ABC be a triangle and P a point in its interior. Prove that*

$$PA + PB + PC < AB + BC + CA.$$

Proof. It seems very plausible that $BP + PC < BA + AC$. Indeed, extending BP to meet AC for the second time at Q, the triangle inequalities in triangles PCQ and ABQ yield

$$BP + PC < BQ + QC < BA + AC.$$

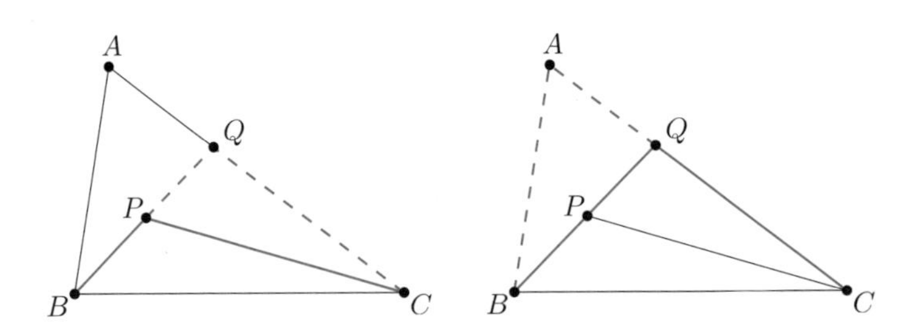

Likewise, we obtain $CP + PA < CB + BA$ and $AP + PB < AC + CB$. Summing these three inequalities and dividing by two we get the result. □

Inequalities with algebraic background

The most common strategy when dealing with inequalities involving elements of a triangle is to rewrite everything in terms of independent variables.

Example 1.26. *Let M, N, P be the midpoints of the sides BC, CA, AB of a triangle ABC and denote by Q, R, S the second intersections of the lines AM, BN, CP with its circumcircle ω. Prove that*

$$\frac{AM}{MQ}+\frac{BN}{NR}+\frac{CP}{PS}\geq 9.$$

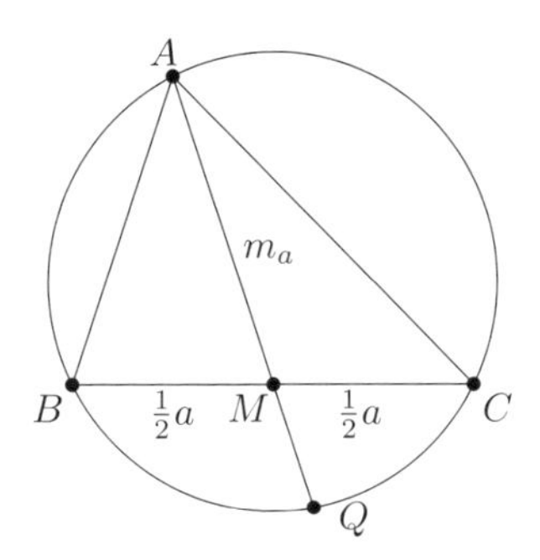

Proof. In order to approach the length MQ we recall Power of a Point and rewrite $MQ=(MB\cdot MC)/MA$. Keeping the median formula $AM^2=\frac{1}{2}(b^2+c^2)-\frac{1}{4}a^2$ (see Corollary 1.24) in mind, we have just expressed everything in terms of the side lengths a, b, c. The rest is straightforward. Indeed, we may write

$$\frac{AM}{MQ}=\frac{AM^2}{MB\cdot MC}=\frac{\frac{1}{2}(b^2+c^2)-\frac{1}{4}a^2}{\left(\frac{1}{2}a\right)^2}=\frac{2(b^2+c^2)}{a^2}-1$$

and similarly for other fractions. Hence it suffices to prove the inequality

$$\frac{b^2}{a^2}+\frac{c^2}{a^2}+\frac{c^2}{b^2}+\frac{a^2}{b^2}+\frac{a^2}{c^2}+\frac{b^2}{c^2}\geq 6$$

which is clearly true as it is the sum of three inequalities of the form $x+1/x\geq 2$ for $x>0$. □

We already know that if a, b, c are the side lengths of a triangle then there exist positive numbers x, y, z such that

$$a=y+z,\qquad b=x+z,\qquad c=x+y$$

(these are precisely the x, y, z used in Proposition 1.26). The advantage of using x, y, z instead of a, b, c is that the former are independent positive real numbers whereas the latter have to satisfy triangle inequalities.

Example 1.27 (IMO 1991). *Prove for each triangle ABC the inequality*

$$\frac{1}{4} < \frac{IA \cdot IB \cdot IC}{l_A l_B l_C} \leq \frac{8}{27},$$

where I is the incenter and l_A, l_B, l_C are the lengths of the angle bisectors of triangle ABC.

Proof. Recalling we know the ratio in which the incenter I divides the angle bisector (see Corollary 1.28), the inequality rewrites as

$$\frac{1}{4} < \frac{b+c}{a+b+c} \cdot \frac{c+a}{a+b+c} \cdot \frac{a+b}{a+b+c} \leq \frac{8}{27}.$$

The second inequality follows immediately from

$$\sqrt[3]{(b+c)(c+a)(a+b)} \leq \frac{2(a+b+c)}{3},$$

which is just AM-GM applied for three terms $a+b$, $b+c$, $c+a$. The first one is however not true for arbitrary a, b, c (try $a=1$, $b=1$, $c=10$) so we have to use the fact that a, b, c are the side lengths of a triangle. The crucial step is to perform the x, y, z substitution which reduces the whole problem into some boring algebra. Denoting $s = x+y+z = \frac{1}{2}(a+b+c)$ it suffices to prove

$$2s^3 < (s+x)(s+y)(s+z),$$

which is true since the right-hand side expands into

$$s^3 + s^2 \underbrace{(x+y+z)}_{=s} + s(xy+yz+zx) + xyz > 2s^3.$$

□

Erdős-Mordell inequality

In the very end of this section we present a famous inequality proposed by Paul Erdős[15] and first proved by L. J. Mordell[16]. Despite its simple statement, the inequality is far from easy to prove (convince yourself!).

Theorem 1.53 (Erdős-Mordell inequality). *Let ABC be a triangle and P a point in its interior. Denote by X, Y, and Z the feet of perpendiculars dropped from P onto BC, CA, and AB, respectively. Then*

$$PA + PB + PC \geq 2(PX + PY + PZ).$$

[15] Paul Erdős (1913–1996) was a Hungarian mathematician. He co-authored over 1500 articles and was perhaps one of the brightest minds of the 20th century.

[16] Louis Joel Mordell (1888–1972) was a British mathematician, known for pioneering research in number theory.

Proof. The key ingredient of the proof is the inequality

$$PA \sin \angle A \geq PY \sin \angle C + PZ \sin \angle B$$

which in fact states that the length of YZ is greater than or equal to its projection onto BC, the latter being equal to the sum of the lengths of the projections of PY and PZ.

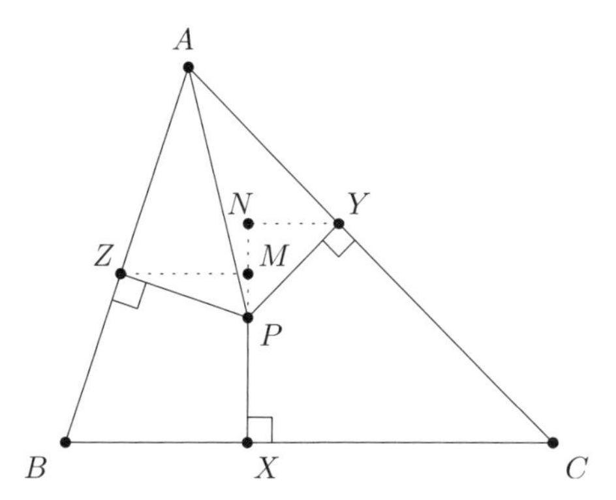

To prove it, note that $AZPY$ is cyclic with diameter AP so the Extended Law of Sines gives $YZ = PA \sin \angle A$. Next, denote the feet of perpendiculars dropped from Z, Y onto PX by M, N, respectively. Since $BZPX$ is cyclic, we have $\angle MPZ = \angle B$ and $ZM = PZ \sin \angle B$. Likewise, $YN = PY \sin \angle C$, so the inequality is indeed equivalent to $YZ \geq YN + MZ$.

To finish the proof of the theorem, let us write the analogous two inequalities for XY and XZ. We obtain

$$PA \geq PY \cdot \frac{\sin \angle C}{\sin \angle A} + PZ \cdot \frac{\sin \angle B}{\sin \angle A},$$
$$PB \geq PZ \cdot \frac{\sin \angle A}{\sin \angle B} + PX \cdot \frac{\sin \angle C}{\sin \angle B},$$
$$PC \geq PX \cdot \frac{\sin \angle B}{\sin \angle C} + PY \cdot \frac{\sin \angle A}{\sin \angle C}.$$

It remains to add these three inequalities and conclude by recalling that for positive x we have $x + 1/x \geq 2$.

The equality occurs if $YZ \parallel BC$, $ZX \parallel AC$, $XY \parallel AB$, and $\sin \angle A = \sin \angle B = \sin \angle C$, i.e. if triangle ABC is equilateral and P is its center. □

Although the power of the Erdős-Mordell inequality might not be apparent at the first glance, the following example demonstrates it entirely.

Example 1.28 (IMO 1991). *Let ABC be a triangle and M a point in its interior. Show that at least one of the angles $\angle MAB$, $\angle MBC$, and $\angle MCA$ is less than or equal to 30°.*

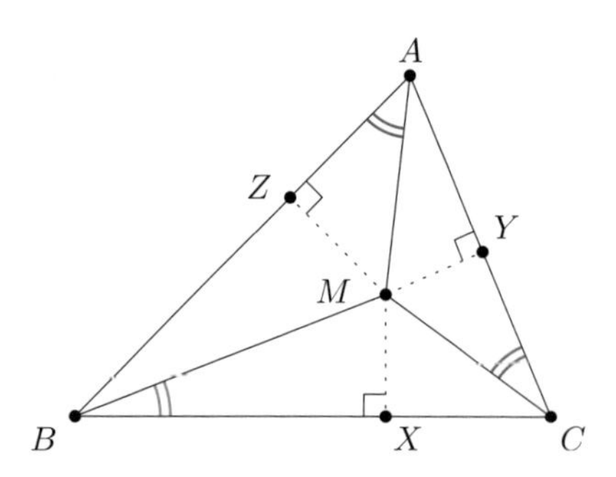

Proof. Denote the feet of perpendiculars dropped from M to the sides BC, CA, AB by X, Y, Z, respectively.

The Erdős-Mordell inequality gives $MB + MC + MA \geq 2(MX + MY + MZ)$, so at least one of the inequalities

$$MB \geq 2MX, \qquad MC \geq 2MY, \qquad MA \geq 2MZ$$

has to hold. Without loss of generality assume $MB \geq 2MX$. Then $\sin MBC = MX/MB \leq \frac{1}{2}$ and thus $\angle MBC \leq 30°$. □

Chapter 2

Introductory Problems

1. Find a polygon and a point in its interior from which no side of the polygon can be seen entirely.

2. Let ABC be a triangle with $AB = AC$ and let K and M be points on the side AB and L a point on the side AC such that $BC = CM = ML = LK = KA$. Find $\angle A$.

3. Let $ABCD$ be a rectangle. Find the locus of points X such that $AX + BX = CX + DX$.

4. Let $a < b < c$ be the sides of a triangle ABC. Prove that $h_b < b$, where h_b is the B-altitude of triangle ABC.

5. On square $ABCD$, point E lies on side AD and point F lies on side BC, so that $BE = EF = FD = 30$. Find the area of square $ABCD$.

6. Let ABC be a triangle with $AB = AC$. Isosceles triangles ABM and ACN with bases AB and AC are erected outside triangle ABC. Prove that the altitudes (possibly extended) $MP \perp AB$, $NQ \perp AC$ and $AA_0 \perp BC$ are concurrent.

7. Squares $ABED$, $BCGF$, $CAIH$ are erected externally from the sides of triangle ABC. Show that triangles AID, BEF, and CGH have equal area.

8. Rhombus $ABCD$ has side length 2 and $\angle B = 120°$. Region $\mathcal{R}$ consists of all points inside the rhombus that are closer to vertex B than any of the other three vertices. What is the area of $\mathcal{R}$?

9. Varignon[1] parallelogram

[1] Pierre Varignon (1654–1722) was a French mathematician.

Let $ABCD$ be a quadrilateral and denote by K, L, M, N the midpoints of the sides AB, BC, CD, DA, respectively.

(a) Prove that $KLMN$ is a parallelogram.

(b) Let P, Q be the midpoints of the diagonals AC, BD, respectively. Prove that $PLQN$ and $PKQM$ are also parallelograms, moreover with the same center.

10. A bus departs from the station S and rides along straight (infinite) road ℓ. Determine the locus of points in the plane from which you can catch the bus if you start running at the time of the departure and you are as fast as the bus.

11. Point P is given inside a circle ω distinct from its center O. Determine the locus of the midpoints of the chords of ω passing through P.

12. Let $ABCD$ be a quadrilateral inscribed in circle ω and let M_a, M_b, M_c, M_d be the midpoints of the arcs AB, BC, CD, DA not containing points C, D, A, and B, respectively. Prove that $M_aM_c \perp M_bM_d$.

13. In rectangle $ABCD$, $AB = 9$ and $BC = 8$. Points E and F lie inside rectangle $ABCD$ so that $EF \parallel AB$, $BE \parallel DF$, $BE = 4$, $DF = 6$, and E is closer to BC than F. Find EF.

14. Distinct points A, B, C lie on a line in this order. Circle ω_1 of radius R passing through A and B intersects circle ω_2 of the same radius R and passing through B and C for the second time at X. Find the locus of X as R varies.

15. Let ABC be a triangle. Equilateral triangles BCD, CAE, ABF are erected outwards from its sides. Show that the circumcircles of these equilateral triangles and the lines AD, BE, CF pass through one point.

16. Triangle ABC has right angle at B, and contains a point P for which $PA = 10$, $PB = 6$, and $\angle APB = \angle BPC = \angle CPA$. Find PC.

17. Points P, Q are given on the sides AB, AD, respectively, of a parallelogram $ABCD$ ($AB > AD$) such that $AP = AQ = x$. Prove that as x varies, the circumcircles of the triangles PQC pass through another fixed point (other than C).

18. In triangle ABC, medians BB_1 and CC_1 are perpendicular. Given that $AC = 19$ and $AB = 22$, find BC.

19. Let ABC be a right triangle with right angle by C and with $CA = 8$, $CB = 6$. Semicircle with diameter CX where $X \in AC$ touches side AB. Find its radius.

20. Four consecutive sides of an equiangular hexagon have lengths 1, 7, 4, and 2. Find the lengths of the remaining two sides.

21. Let ABC be a triangle with $\angle A = 60°$ and denote its incenter by I. Lines BI, CI intersect the opposite sides at E, F, respectively. Prove that $IE = IF$.

22. In quadrilateral $ABCD$ let $BC = 8$, $CD = 12$, $AD = 10$, and $\angle A = \angle B = 60°$. Find the distance AB.

23. Let $ABCD$ be a convex quadrilateral. Find point X for which the sum of distances to its vertices is minimal.

24. Circles ω_1, ω_2 intersect at points A and B. An arbitrary line passing through B intersects ω_1 for the second time at K (outside ω_2) and ω_2 at L (outside ω_1).

    (a) Prove that all possible triangles AKL are similar to each other.

    (b) Let the tangents at points K and L to the respective circles intersect at P. Prove that $KPLA$ is cyclic.

25. Let $ABCD$ be a quadrilateral with $AB \parallel CD$. If $\angle ADB + \angle DBC = 180°$, prove that

$$\frac{AB}{CD} = \frac{AD}{BC}.$$

26. On side BC of triangle ABC an arbitrary point D is selected. The tangent in D to the circumcircle of triangle ABD meets AC at point B_1. Point C_1 is defined analogously. Prove that $B_1C_1 \parallel BC$.

27. Conway's[2] circle.

    Let ABC be a triangle and denote by A_1, A_2 the points on the rays opposite to AB, AC, respectively, satisfying $AA_1 = AA_2 = BC$. Define points B_1, B_2, C_1, C_2 analogously. Prove that points A_1, A_2, B_1, B_2, C_1, C_2 lie on a single circle.

28. Let ABC be an acute triangle. Prove that $h_a > \frac{1}{2}(b + c - a)$, where h_a is the length of A-altitude in triangle ABC.

[2]John Horton Conway (1936) is a contemporary British mathematician known for many delightful discoveries both in recreational and research mathematics.

29. Let ABC be a triangle. Find the locus of points X $(X \neq A)$ for which the triangles AXB and AXC have equal area.

30. Let $ABCD$ be a quadrilateral with perpendicular diagonals inscribed in a circle with radius R. Prove that

$$AB^2 + BC^2 + CD^2 + DA^2 = 8R^2.$$

31. A trapezoid $ABCD$ has AB parallel to CD. The external bisectors of $\angle A$ and $\angle D$ meet at P, and the external bisectors of $\angle B$ and $\angle C$ meet at Q. Show that PQ is half the perimeter of $ABCD$.

32. In triangle ABC, $11 \cdot AB = 20 \cdot AC$. The angle bisector of $\angle A$ intersects BC at point D, and point M is the midpoint of AD. Let P be the point of intersection of AC and BM. Find CP/PA.

33. A variable segment BC of fixed length d moves such that its endpoints remain on the fixed rays AU, AV. Prove that the circumcircles of all possible triangles ABC are all tangent to a fixed circle.

34. Let ABC be a right triangle with $\angle A = 90^\circ$ and altitude AD. Let r, s, t be the inradii of triangles ABC, ADB, and ADC, respectively. Show that $r + s + t = AD$.

35. In triangle ABC, $BC = 125$, $CA = 120$, and $AB = 117$. The angle bisector of angle B intersects CA at point K, and the angle bisector of angle C intersects AB at point L. Let M and N be the feet of the perpendiculars from A to CL and BK, respectively. Find MN.

36. Let $ABCD$ and $AB'C'D'$ be parallelograms such that B' lies on the segment BC and D lies on the segment $C'D'$. Show that their areas are equal.

37. In triangle ABC there is a point F on the side AB such that $\angle FAC = \angle FCB$ and $AF = BC$. Further, BE is the internal angle bisector of $\angle B$ with $E \in AC$. Show that $EF \parallel BC$.

38. Let I be the incenter of triangle ABC. Prove that

$$\frac{AI^2}{bc} + \frac{BI^2}{ca} + \frac{CI^2}{ab} = 1.$$

39. In parallelogram $ABCD$ with $\angle BAD > 90^\circ$, show that the circle passing through the projections of C onto AB, BD, and DA, respectively, passes through the center of the parallelogram.

40. Let $ABCD$ be a cyclic quadrilateral. Let P be the point on the ray AD such that $AP = BC$ and let Q be the point on the ray AB such that $AQ = CD$. Prove that the line AC cuts PQ at its midpoint.

41. Let $ABCDE$ be a convex pentagon such that $AB + CD = BC + DE$ and a circle ω with center O on the side AE is tangent to the sides AB, BC, CD and DE at points P, Q, R and S, respectively. Prove that the lines PS and AE are parallel.

42. Let P be a point inside acute-angled triangle ABC with $\angle BPC = 180 - \angle A$. Denote by A_1, B_1, C_1 its reflections over the sides BC, CA, AB, respectively. Prove that the points A, A_1, B_1, C_1 are concyclic.

43. Triangle KLM lies inside triangle ABC so that points K, L, M lie on the segments CL, AM, BK, respectively. Prove that the circumcircles of the triangles ABM, BCK, CAL pass through a common point.

44. Let the pentagon $ABCDE$ inscribed in circle ω satisfy $BA = BC$. The line joining $P = BE \cap AD$ and $Q = CE \cap BD$ intersects ω at points X, Y. Prove that $BX = BY$.

45. In the convex pentagon $ABCDE$ all interior angles have the same measure. Prove that the perpendicular bisector of segment EA, the perpendicular bisector of segment BC and the angle bisector of $\angle CDE$ intersect at one point.

46. Let ω_1, ω_2 be two circles. One of their common external tangents is tangent to ω_1 at A, the second one is tangent to ω_2 at D. Line AD intersects the circles ω_1, ω_2 for the second time at B, C, respectively. Prove that $AB = CD$.

47. Triangle ABC has $AB = 13$, $BC = 14$, and $CA = 15$. The points D, E, and F are the midpoints of BC, CA, and AB, respectively. Let $X \neq D$ be the intersection of the circumcircles of triangles BDF and CDE. What is $XA + XB + XC$?

48. Let $ABCD$ be a quadrilateral with segments BC and AD equal and AB not parallel to CD. Denote by M, N the midpoints of BC and AD, respectively. Prove that the perpendicular bisectors of AB, MN, and CD pass through a common point.

49. Carnot's[3] Theorem.

[3]Lazare Nicolas Marguerite Carnot (1753–1823) was an amateur mathematician and French minister of war during the French revolutionary wars.

Let X, Y, and Z lie on the sides BC, CA, AB, respectively, of a triangle ABC. Show that the perpendiculars from X, Y, Z to the respective triangle sides meet at one point if and only if

$$BX^2 + CY^2 + AZ^2 = CX^2 + AY^2 + BZ^2.$$

50. In a given pentagon $ABCDE$, triangles ABC, BCD, CDE, DEA and EAB all have the same area. The lines AC and AD intersect BE at points M and N. Prove that $BM = EN$.

51. Let ABC be a non-right triangle with orthocenter H and let M, N be points on its sides AB and AC. Prove that the common chord of circles with diameters CM and BN passes through H.

52. Let fixed points A, Z, B lie on a line ℓ in this order such that $ZA \neq ZB$. A variable point $X \notin \ell$ and a variable point Y on the segment XZ are chosen. Let $D = BY \cap AX$ and $E = AY \cap BX$. Prove that all lines DE pass through a fixed point.

53. Let ω_1 and ω_2 be two circles centered at distinct points O_1 and O_2 and with radii r_1, r_2, respectively.

    (a) Find the locus of points X for which $p(X, \omega_1) - p(X, \omega_2)$ is constant.

    (b) Find the locus of points X for which $p(X, \omega_1) + p(X, \omega_2)$ is constant.

Chapter 3

Advanced Problems

1. On the sides AB and AD of the rhombus $ABCD$ consider the points E and F such that $AE = DF$. Let $BC \cap DE = P$ and $CD \cap BF = Q$. Prove that points P, A, and Q are collinear.

2. Let $ABCD$ be a parallelogram such that the triangle ABD is acute and has orthocenter H. The line through H parallel to AB cuts AD and BC at Q and P, respectively, while the line through H parallel to BC cuts AB and CD at R and S, respectively. Prove that the points P, Q, R, S lie on the same circle.

3. Let ABC be an acute-angled triangle. Let D and E be points on the sides AB and AC such that B, C, D, and E lie on the same circle. Further, suppose the circle through D, E, and A intersects the side BC in two points X and Y. Show that the midpoint of XY is the foot of the altitude from A to BC.

4. Point B lies on a line which is tangent to circle ω at point A. The line segment AB is rotated about the center of the circle by some angle to form segment $A'B'$. Prove that the line AA' bisects the segment BB'.

5. Let ω_1 and ω_2 be concentric circles, with ω_2 in the interior of ω_1. From a point A on ω_1 draw the tangent AB to ω_2 ($B \in \omega_2$). Let C be the second point of intersection of AB and ω_1, and let D be the midpoint of AB. A line passing through A intersects ω_2 at E and F in such a way that the perpendicular bisectors of DE and CF intersect at a point M on AB. Find the ratio AM/MC.

6. Let M be a point inside triangle ABC such that

$$AM \cdot BC + BM \cdot AC + CM \cdot AB = 4[ABC].$$

Show that M is the orthocenter of triangle ABC.

7. Let ABC be a triangle. Prove that lines joining midpoints of the sides with midpoints of the corresponding altitudes pass through a single point.

8. Let $ABCD$ be a convex quadrilateral such that $\angle ADB = \angle BDC$. Suppose that a point E on the side AD satisfies the equality
$$AE \cdot ED + BE^2 = CD \cdot AE.$$
Show that $\angle EBA = \angle DCB$.

9. Let ABC be a triangle with $\angle A = 90°$. Denote its incenter by I and let $D = BI \cap AC$ and $E = CI \cap AB$. Determine whether or not it is possible for segments AB, AC, BI, ID, CI, IE to all have integer lengths.

10. Let A and B be two fixed points inside of the fixed circle ω symmetric with respect to its center O. If points M and N vary on ω in the same half-plane with respect to AB, so that $AM \parallel BN$, prove that $AM \cdot BN$ is constant.

11. In a trapezoid $ABCD$, the segment connecting the midpoints M, N of the bases AB, CD, respectively, has length 4, and the diagonals have lengths $AC = 6$ and $BD = 8$ Find the area of the trapezoid.

12. In triangle ABC, let AP, BQ, CR be concurrent cevians. Let the circumcircle of triangle PQR intersect the sides BC, CA, AB for the second time at X, Y, Z, respectively. Prove that AX, BY, CZ are concurrent.

13. A quadrilateral $ABCD$ is inscribed in a circle ω. The tangent to ω at B intersects the ray DC at K, and the tangent to ω at C intersects the ray AB at M. Prove that if $BM = BA$ and $CK = CD$, then $ABCD$ is a trapezoid.

14. Let $ABCD$ be a parallelogram and M, N points on its sides AB, AD such that $\angle MCB = \angle DCN$. Let P, Q, R, and S be the midpoints of the segments AB, AD, NB, and MD, respectively. Show that P, Q, R, and S are concyclic.

15. Diagonals of non-isosceles trapezoid $ABCD$ intersect at P. Let A_1 be the second intersection of the circumcircle of triangle BCD and AP. Points B_1, C_1, D_1 are defined in a similar way. Prove that $A_1B_1C_1D_1$ is also a trapezoid.

16. Let ω be a circle with center O and radius r and A a point different from O. Find the locus of circumcenters of the triangles ABC for which BC is a diameter of ω.

17. Let $ABCD$ be quadrilateral such that

$$\angle ADB + \angle ACB = 90^\circ \quad \text{and} \quad \angle DBC + 2\angle DBA = 180^\circ.$$

Show that

$$(DB + BC)^2 = AD^2 + AC^2.$$

18. We are given a triangle ABC such that $AB = AC$. There is a point D lying on the segment BC, such that $BD < DC$. Point E is symmetrical to B with respect to AD. Prove that

$$\frac{AB}{AD} = \frac{CE}{CD - BD}.$$

19. Let P be a point on the side BC of triangle ABC. Perpendicular bisectors of the sides AB and AC meet the segment AP at points D and E, respectively. The line parallel to AB passing through D intersects the tangent to the circumcircle ω of triangle ABC through B at point M. Similarly, the line parallel to AC passing through E intersects the tangent to ω through C at point N. Prove that MN is tangent to ω.

20. In an acute triangle ABC a semicircle ω centered on the side BC is tangent to the sides AB and AC at points F and E, respectively. If X is the intersection of BE and CF, show that $AX \perp BC$.

21. Let $ABCD$ be a convex quadrilateral and X a point in its interior. Denote by ω_A the circle tangent to the sides AB and AD and passing through X. Define circles ω_B, ω_C, and ω_D similarly. Given that all these circles have equal radii, show that $ABCD$ is cyclic.

22. In triangle ABC, let AP, BQ, CR be concurrent cevians. Denote by X, Y, Z the midpoints of segments QR, RP, PQ, respectively. Prove that the lines AX, BY, CZ are concurrent.

23. Given a triangle ABC, let P and Q be points on segments AB and AC, respectively, such that $AP = AQ$. Let S and R be distinct points on segment BC such that S lies between B and R, $\angle BPS = \angle PRS$, and $\angle CQR = \angle QSR$. Prove that P, Q, R, S are concyclic.

24. Segment AT is tangent to circle ω at T. A line parallel to AT intersects ω at B, C (with $AB < AC$). Lines AB, AC intersect ω for the second time at P, Q. Prove that line PQ bisects segment AT.

25. Diagonals AC and BD of a cyclic quadrilateral $ABCD$ meet at P. Let the circumcenters of $ABCD$, ABP, BCP, CDP, and DAP be O, O_1, O_2, O_3, and O_4, respectively. Prove that OP, O_1O_3, and O_2O_4 are concurrent.

26. Triangle ABC has $BC = 20$. The incircle of the triangle evenly trisects the median AD at points E and F. Find the area of the triangle.

27. Let P be a point inside an equilateral triangle ABC. Let the lines AP, BP, CP meet the sides BC, CA, AB at the points A_1, B_1, C_1, respectively. Prove that

$$A_1B_1 \cdot B_1C_1 \cdot C_1A_1 \geq A_1B \cdot B_1C \cdot C_1A.$$

28. Let P and Q be isogonal conjugates[1] with respect to the triangle ABC. Show that the six feet of perpendiculars from P and Q to the sides of triangle ABC lie on one circle.

29. The incircle of triangle ABC is tangent to its sides BC, CA, AB at points D, E, F, respectively. The excircles of triangle ABC are tangent to the corresponding sides of triangle ABC at points T, U, V. Show that triangles DEF and TUV have the same area.

30. Let H be the orthocenter of an acute-angled triangle ABC. The circle Γ_A centered at the midpoint of BC and passing through H intersects the sideline BC at points A_1 and A_2. Similarly, define the points B_1, B_2, C_1, and C_2. Prove that six points A_1, A_2, B_1, B_2, C_1, and C_2 are concyclic.

31. Distinct points A, B are given in the plane. Determine the locus of points C such that in triangle ABC the length of A-altitude is the same as the length of B-median.

32. Let ABC be an acute triangle with altitudes BB_0 and CC_0. Point P is given such that the line PB is tangent to the circumcircle of triangle PAC_0 and the line PC is tangent to the circumcircle of triangle PAB_0. Prove that AP is perpendicular to BC.

33. Let ABC be a triangle. Point O in its interior satisfies $\angle OBA = \angle OAC$, $\angle BAO = \angle OCB$, and $\angle BOC = 90°$. Find AC/OC.

34. Let $ABCD$ be a cyclic quadrilateral ($AB \neq CD$). Quadrilaterals $AKDL$ and $CMBN$ are rhombi with equal sides. Prove that points K, L, M, N lie on a single circle.

[1] For explanation see Theorem 1.46.

35. Let ABC be a triangle with inradius r and let ω be a circle of radius $a < r$ inscribed in angle BAC. Tangents from B and C to ω (different from the triangle sides) intersect at point X. Show that the incircle of triangle BCX is tangent to the incircle of triangle ABC.

36. Let $ABCD$ be a cyclic quadrilateral. Let P, Q, R be the feet of the perpendiculars from D to the lines BC, CA, AB, respectively. Show that $PQ = QR$ if and only if the bisectors of $\angle ABC$ and $\angle ADC$ are concurrent with AC.

37. Let X be a point on the circumcircle of a cyclic quadrilateral $ABCD$. Denote by E, F, G, and H the projections of X onto lines AB, BC, CD, DA, respectively. Prove that

$$BE \cdot CF \cdot DG \cdot AH = AE \cdot BF \cdot CG \cdot DH.$$

38. Newton-Gauss[2] line.

    Let $ABCD$ be a convex quadrilateral. Denote by Q the intersection of AD and BC and by R the intersection of AB and CD. Let X, Y, and Z be the midpoints of AC, BD, and QR, respectively. Prove that X, Y, and Z lie on a single line.

39. In acute triangle ABC let A_1, B_1 be the points of tangency of A-excircle with BC and B-excircle with AC, respectively. Let H_1, H_2 be the orthocenters of triangles CAA_1 and CBB_1, respectively. Prove that H_1H_2 is perpendicular to the angle bisector of $\angle ACB$.

40. A circle ω with center O is internally tangent to two circles in its interior at points S and T which are not diametrically opposite. Suppose the two circles intersect at M and N with N closer to ST. Show that $OM \perp MN$ if and only if S, N, T are collinear.

41. Let $ABCD$ be a quadrilateral with an inscribed circle ω and let the points of tangency of the incircle with sides AB, BC, CD, DA be K, L, M, N, respectively. Prove that the lines AC, BD, KM, and LN are concurrent.

42. Orthologic triangles.

    Let ABC and $A'B'C'$ be two triangles in plane. Show that the perpendiculars from A' to BC, from B' to CA and from C' to AB (denote their feet by X, Y, and Z, respectively) are concurrent if and only if the perpendiculars from A to $B'C'$, from B to $C'A'$, and from C to $A'B'$ are concurrent.

---

[2]Johann Carl Friedrich Gauss (1777–1855) was a German mathematician and physicist.

43. Let ABC be a triangle with medians m_a, m_b, m_c and circumradius R. Prove that
$$\frac{b^2+c^2}{m_a}+\frac{c^2+a^2}{m_b}+\frac{a^2+b^2}{m_c}\leq 12R.$$

44. Show that in acute triangle ABC we have $r+R\leq h$, where r, R, and h are the inradius, circumradius and the longest altitude, respectively.

45. Let P be a point in the plane of triangle ABC, and ℓ a line passing through P. Let A', B', C' be the points where the reflections of lines PA, PB, PC with respect to ℓ intersect lines BC, AC, AB respectively. Prove that A', B', C' are collinear.

46. Let BC be the longest side of a scalene triangle ABC. Point K on the ray CA satisfies $KC = BC$. Similarly, point L on the ray BA satisfies $BL = BC$. Prove that KL is perpendicular to OI where O, I denote the circumcenter and the incenter of triangle ABC, respectively.

47. Let D be an arbitrary point on the side BC of a given triangle ABC and let E be the intersection of AD and the second external common tangent of the incircles of triangles ABD and ACD. As D assumes all positions between B and C, prove that the point E traces an arc of a circle.

48. Let ABC be a triangle with circumcenter O. The points P and Q are interior points of the sides CA and AB, respectively. Let K, L and M be the midpoints of the segments BP, CQ and PQ, respectively, and let Γ be the circle passing through K, L, and M. Suppose that the line PQ is tangent to the circle Γ. Prove that $OP = OQ$.

49. Let ABC be a non-right triangle. A circle ω passing through B and C intersects the sides AB and AC again at C' and B', respectively. Prove that BB', CC' and HH' are concurrent, where H and H' are the orthocenters of triangles ABC and $AB'C'$, respectively.

50. Let P be a point in the interior of triangle ABC with circumradius R. Prove that
$$\frac{AP}{a^2}+\frac{BP}{b^2}+\frac{CP}{c^2}\geq\frac{1}{R}.$$

51. Let $AXYZB$ be a convex pentagon inscribed in a semicircle of diameter AB. Denote by P, Q, R, S the feet of the perpendiculars from Y onto lines AX, BX, AZ, BZ, respectively. Prove that the acute angle formed by lines PQ and RS is half the size of $\angle ZOX$, where O is the midpoint of the segment AB.

52. Let PAB and PCD be triangles such that $PA = PB$, $PC = PD$, and triads of points P, A, C and B, P, D are both collinear in this order. A circle ω_1 passing through A and C intersects a circle ω_2 passing through B and D at distinct points X, Y. Prove that the circumcenter of the triangle PXY is the midpoint of the segment formed by the centers O_1, O_2 of ω_1, ω_2, respectively.

53. Let ABC be a triangle with $\angle BCA = 90°$, and let D be the foot of the altitude from C. Let X be a point in the interior of the segment CD. Let K be the point on the segment AX such that $BK = BC$. Similarly, let L be the point on the segment BX such that $AL = AC$. Let M be the point of intersection of AL and BK.

    Show that $MK = ML$.

Chapter 4

Solutions to Introductory Problems

1. Find a polygon and a point in its interior from which no side of the polygon can be seen entirely.

   Solution. From the many possible solutions we offer two which have similar flavor.

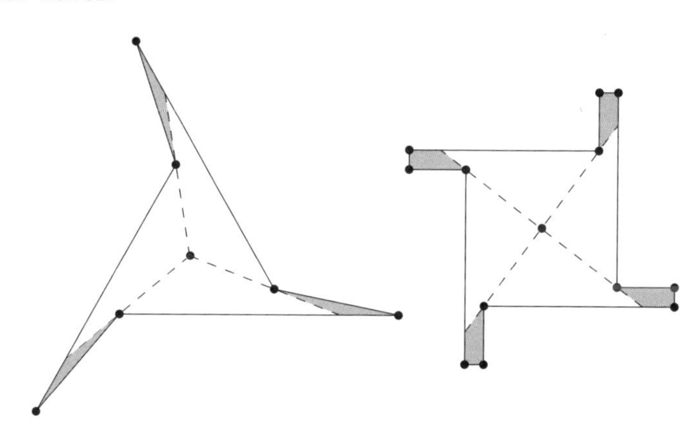

2. Let ABC be a triangle with $AB = AC$ and let K and M be points on the side AB and L a point on the side AC such that $BC = CM = ML = LK = KA$. Find $\angle A$.

   Solution. Denote $\angle A$ by α. The only ingredient in this proof is that we interpret equal distances as equal angles. We have isosceles triangles BCM, CML, MLK, and LKA. Starting from triangle LKA we learn that $\angle ALK = \alpha$ and from external angle in triangle KLA we have $\angle MKL = 2\alpha$. In the same way, we obtain $\angle LMK = 2\alpha$ (triangle MLK is isosceles) and $\angle MLC = 3\alpha$ (external angle in triangle ALM). Finally, we apply this step one more time to get $\angle CBA = \angle BMC = 3\alpha + \alpha = 4\alpha$. Since triangle ABC itself is isosceles we have

$$180^\circ = \angle ACB + \angle BCA + \angle BAC = 4\alpha + 4\alpha + \alpha = 9\alpha.$$

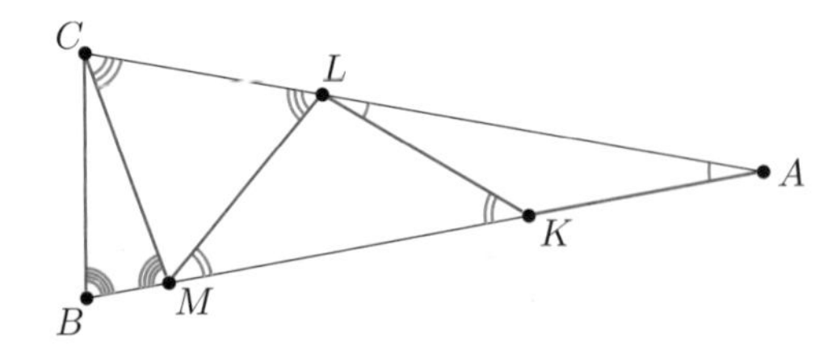

Thus the answer is $\alpha = 20^\circ$.

3. Let $ABCD$ be a rectangle. Find the locus of points X such that $AX + BX = CX + DX$.

   Solution. Draw the common perpendicular bisector ℓ of BC and AD and assume it is horizontal with AB below ℓ. Points X on ℓ obviously satisfy the desired condition since $BX = CX$ and $AX = DX$. Points X' which are above ℓ have $BX' > CX'$ and $AX' > DX'$, therefore they don't fulfil the condition. For analogous reason we can also exclude points below ℓ. Thus the locus is exactly the line ℓ.

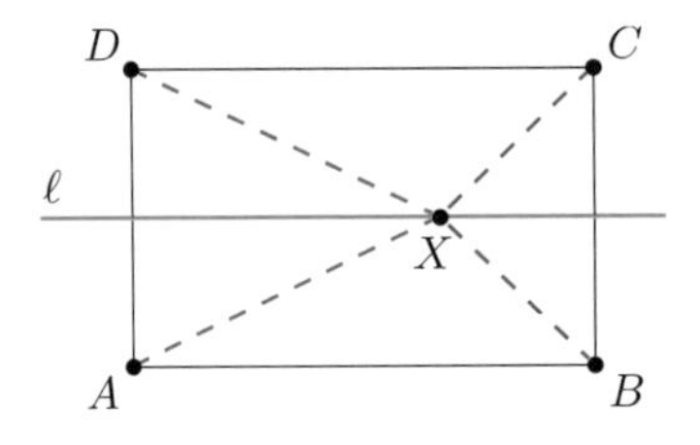

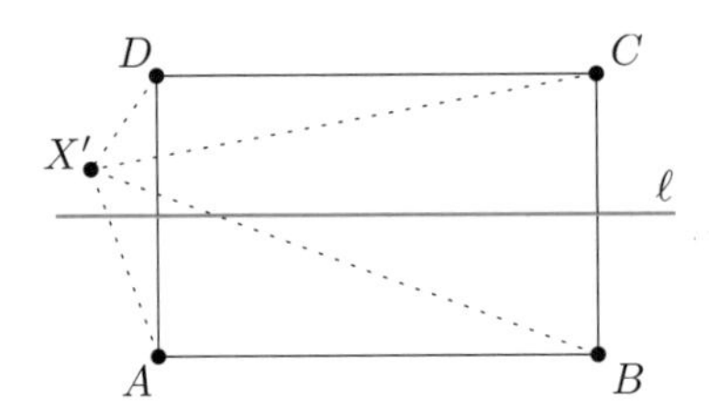

4. Let $a < b < c$ be the sides of a triangle ABC. Prove that $h_b < b$, where h_b is the B-altitude of triangle ABC.

   Proof. Since the B-altitude is the shortest distance from B to line AC, we certainly have $h_b \leq a$ and since $a < b$ the result follows.

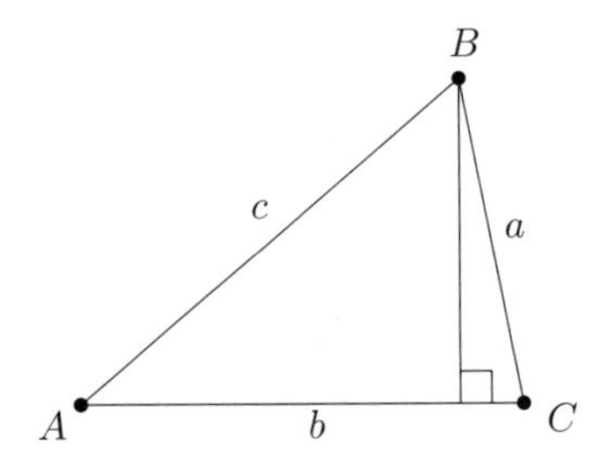

5. [AIME 2011] On square $ABCD$, point E lies on side AD and point F lies on side BC, so that $BE = EF = FD = 30$. Find the area of square $ABCD$.

Solution. If we place AB horizontally and draw horizontal lines also through points E and F, we see that we have divided the square into six pairwise congruent triangles (HL).

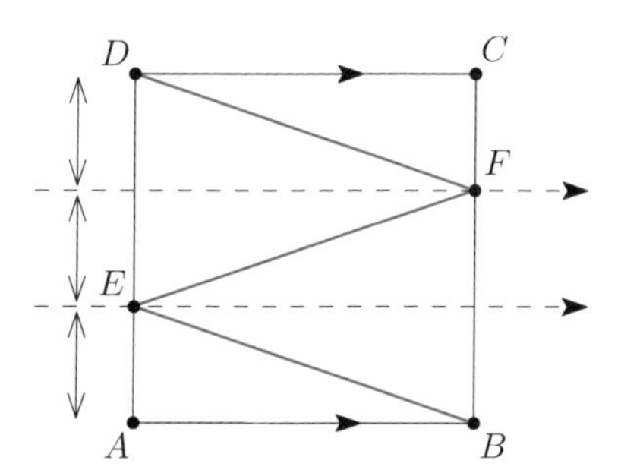

Thus, $AE = \frac{1}{3}a$, where a is the side length of the square. From the Pythagorean Theorem we learn

$$30^2 = BE^2 = a^2 + \left(\frac{1}{3}a\right)^2 = \frac{10}{9}a^2,$$

or equivalently $a^2 = 810$, which is our final answer.

6. Let ABC be a triangle with $AB = AC$. Isosceles triangles ABM and ACN with bases AB and AC are erected outside triangle ABC. Prove that the altitudes (possibly extended) $MP \perp AB$, $NQ \perp AC$ and $AA_0 \perp BC$ are concurrent.

Proof. In isosceles triangle ABM the altitude MP coincides with the perpendicular bisector of AB. Similarly, NQ is the perpendicular bisector of AC and AA_0 is the perpendicular bisector of BC. Since the altitudes are in fact the perpendicular bisectors of triangle ABC, they concur at its circumcenter O.

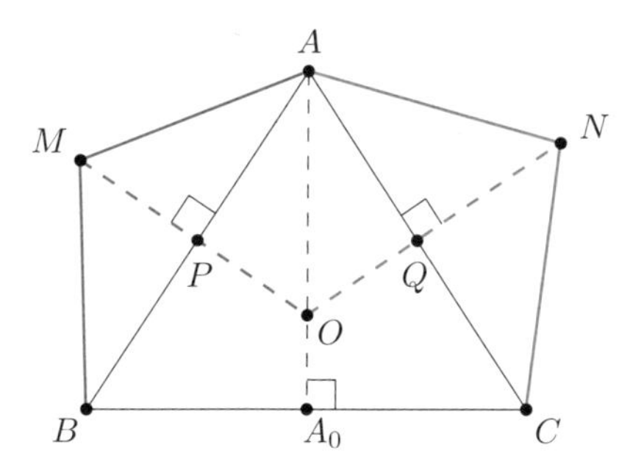

7. Squares $ABED$, $BCGF$, $CAIH$ are erected externally from the sides of triangle ABC. Show that triangles AID, BEF, and CGH have equal area.

   Proof. We turn our attention to triangle DAI. We have $AD = AB$ and $AI = AC$ and also

$$\angle IAD = 360^\circ - 90^\circ - 90^\circ - \angle BAC = 180^\circ - \angle BAC.$$

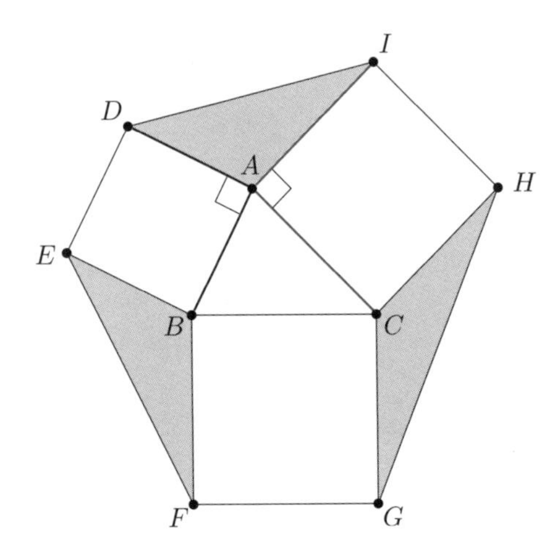

   Thus we can compute the area K_A of the triangle DAI as

$$K_A = \frac{1}{2} AD \cdot AI \cdot \sin \angle IAD = \frac{1}{2} AB \cdot AC \cdot \sin(180^\circ - \angle BAC) =$$
$$= \frac{1}{2} AB \cdot AC \cdot \sin \angle BAC$$

   which is exactly the area of triangle ABC! Thus by symmetry all three triangles have area equal to that of triangle ABC and the conclusion follows.

8. [AMC12 2011] Rhombus $ABCD$ has side length 2 and $\angle B = 120^\circ$. Region $\mathcal{R}$ consists of all points inside the rhombus that are closer to vertex B than any of the other three vertices. What is the area of $\mathcal{R}$?

   Solution. First, recall that the locus of points which are closer to point X than to point Y is a half-plane with the perpendicular bisector of XY as borderline. In this case the borderlines of $\mathcal{R}$ will be the perpendicular bisectors of BA, BC, and BD.

   Note that triangles ABD and BCD are both equilateral as BD bisects the congruent angles ABC and CDA. Now observe that if we connect midpoint of each side of triangle ABD with its center, we divide the

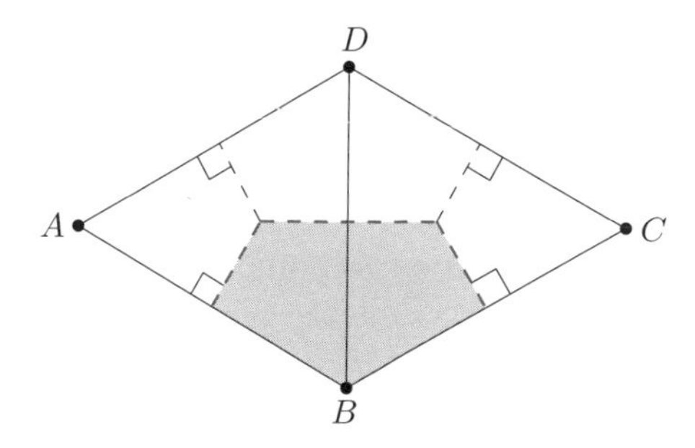

triangle into three congruent regions, one of which is exactly one half of $\mathcal{R}$. Thus $\mathcal{R}$ takes up exactly one third of each of triangles ABD and BCD. Calculating the area of equilateral triangle we find the desired area K as

$$K = \frac{1}{3}([ABD] + [BCD]) = \frac{2}{3}[ABD] = \frac{2}{3} \cdot \frac{\sqrt{3}}{4} BD^2 = \frac{2}{3}\sqrt{3}.$$

9. Varignon[1] parallelogram

   Let $ABCD$ be a quadrilateral and denote by K, L, M, N the midpoints of the sides AB, BC, CD, DA, respectively.

   (a) Prove that $KLMN$ is a parallelogram.

   (b) Let P, Q be the midpoints of the diagonals AC, BD, respectively. Prove that $PLQN$ and $PKQM$ are also parallelograms, moreover with the same center.

   Proof.

   (a) The key is to realize that both KL and NM are midlines in some triangles. Namely, in triangles ABC and ADC. Thus $KL \parallel AC \parallel NM$ and also $KL = \frac{1}{2}AC = NM$. This ensures that $KLMN$ is a parallelogram.

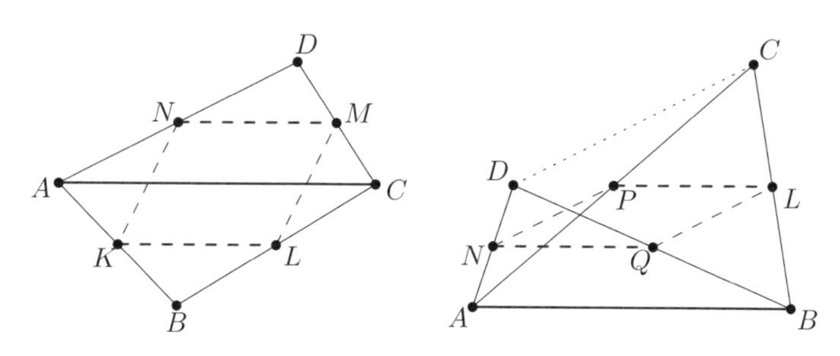

[1] Pierre Varignon (1654–1722) was a French mathematician.

(b) We apply the same idea. Segments PL and NQ are midlines in triangles ACB and ADB, respectively, therefore again $PL \parallel AB \parallel NQ$ and $PL = \frac{1}{2}AB = NQ$ and $PLQN$ is a parallelogram. Also, since diagonals in a parallelogram bisect each other, the center of this parallelogram is the midpoint of NL, just like in (a).

For quadrilateral $PKQM$ we proceed analogously.

10. A bus departs from the station S and rides along straight (infinite) road ℓ. Determine the locus of points in the plane from which you can catch the bus if you start running at the time of the departure and you are as fast as the bus.

Solution. We can clearly catch the bus from S. For any other point X catching the bus from X is equivalent to finding a point on the ray ℓ (let's assume it's horizontal and points to the right) which is closer to X than to S (or equidistant from them). This happens if and only if the perpendicular bisector of XS intersects ℓ or in other words if and only if the angle between ℓ and XS is less than $90°$.

Thus, denoting by m the perpendicular to ℓ through S, the answer is "point S and the half-plane consisting of all the points to the right of m".

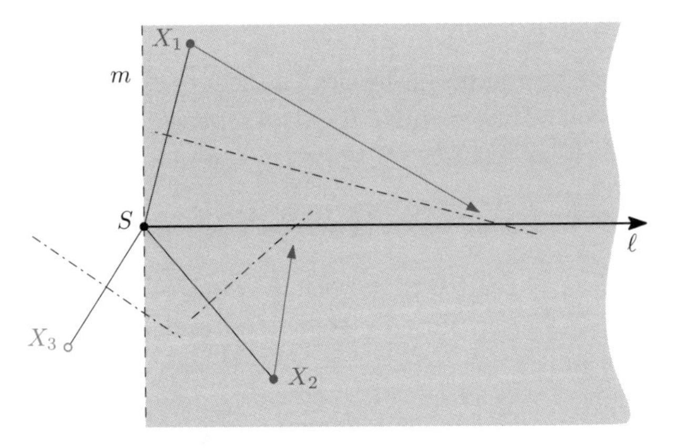

11. Point P is given inside a circle ω distinct from its center O. Determine the locus of the midpoints of the chords of ω passing through P.

Solution. First observe that if we draw a chord through both P and O, then its midpoint is O. Now consider some other chord ℓ and denote its midpoint by X. Clearly, as O is the center of the circle, it is equidistant from the endpoints of ℓ and thus it lies on its perpendicular bisector.

In other words $OX \perp \ell$. If $X \neq P$, this means $\angle OXP = 90°$. Thus, we are restricted to a circle with diameter OP and it is easy to verify

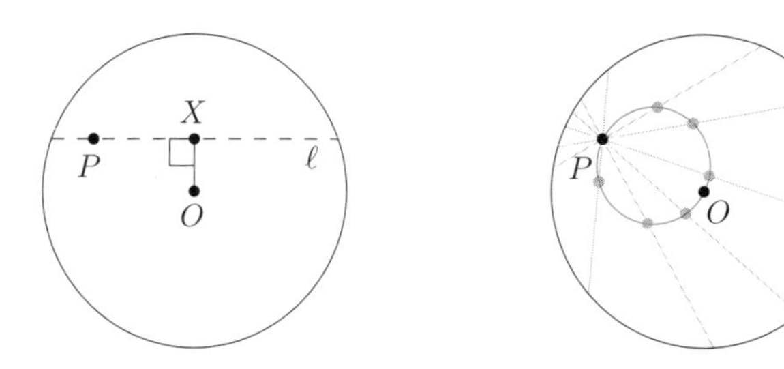

that all points of this circle (including P) are indeed midpoints of some chord passing through P.

12. Let $ABCD$ be a quadrilateral inscribed in circle ω and let M_a, M_b, M_c, M_d be the midpoints of the arcs AB, BC, CD, DA not containing points C, D, A, and B, respectively. Prove that $M_aM_c \perp M_bM_d$.

    Proof. We divide the circle into the four arcs AB, BC, CD, and DA and label the corresponding inscribed angles as α, β, γ, δ. Now we calculate the angle between chords M_aM_c and M_bM_d as sum of the corresponding inscribed angles (see Corollary 1.32).

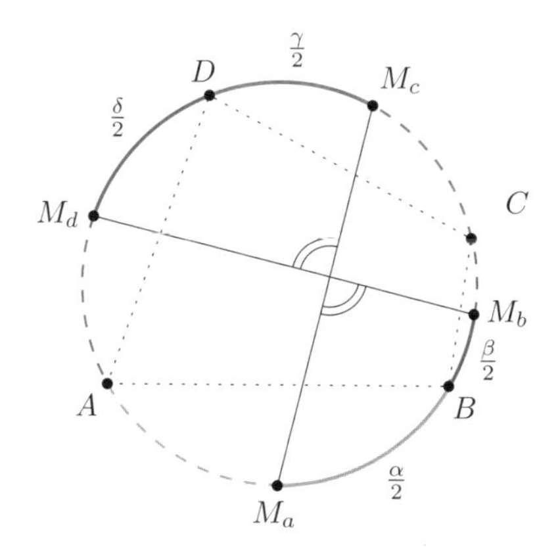

We obtain

$$\angle(M_aM_c, M_bM_d) = \left(\frac{\alpha}{2} + \frac{\beta}{2}\right) + \left(\frac{\gamma}{2} + \frac{\delta}{2}\right) = 90^\circ,$$

where we used the notion of directed angles.

13. [based on AIME 2011] In rectangle $ABCD$, $AB = 9$ and $BC = 8$. Points E and F lie inside rectangle $ABCD$ so that $EF \parallel AB$, $BE \parallel DF$, $BE = 4$, $DF = 6$, and E is closer to BC than F. Find EF.

    Solution. We draw our diagram so that AB is horizontal and draw vertical lines through points E and F. In order to connect the segments DF and EB, we simply cut away the vertical strip.

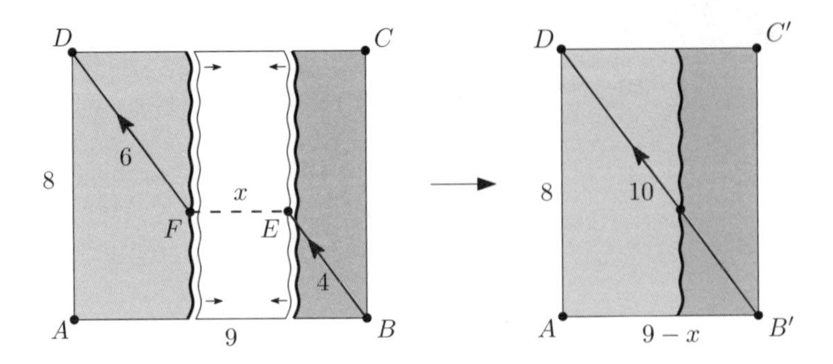

If we denote EF by x, Pythagorean theorem in the newly formed right triangle gives

$$(9-x)^2 + 8^2 = 10^2 \quad \text{or} \quad (9-x)^2 = 6^2.$$

Since $x < 9$ (EF lies inside $ABCD$) the solution is $x = 3$.

14. Distinct points A, B, C lie on a line in this order. Circle ω_1 of radius R passing through A and B intersects circle ω_2 of the same radius R and passing through B and C for the second time at X. Find the locus of X as R varies.

First Solution. Draw the common chord XB. Since the circles ω_1 and ω_2 are congruent, the inscribed angles corresponding to the same arc XB are the same. In other words, $\angle XAB = \angle XCB$ so the triangle AXC is isosceles. Hence X lies on the perpendicular bisector of AC.

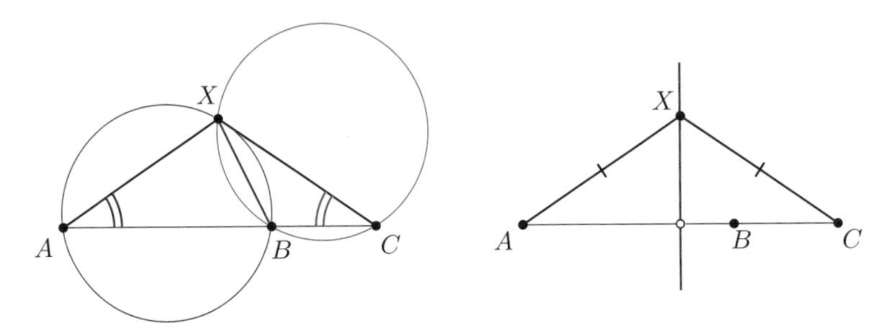

On the other hand, any such point X' distinct from the midpoint of AC can be attained as the circumcircles of $X'BC$ and $X'AB$ have the same radii. Thus the locus is the perpendicular bisector of AC without the midpoint of AC.

Second Solution. Using the Extended Law of Sines in triangles AXB, BXC we obtain

$$\frac{XA}{\sin \angle XBA} = 2R = \frac{XC}{\sin XBC}.$$

As the angles XBA and XBC are supplementary, their sines are the same. Hence $XA = XC$ and we continue as in the first solution.

15. Let ABC be a triangle. Equilateral triangles BCD, CAE, ABF are erected outwards from its sides. Show that the circumcircles of these equilateral triangles and the lines AD, BE, CF pass through one point.

Proof. Let P be the second intersection of the circumcircles of triangles ABF and ACE.

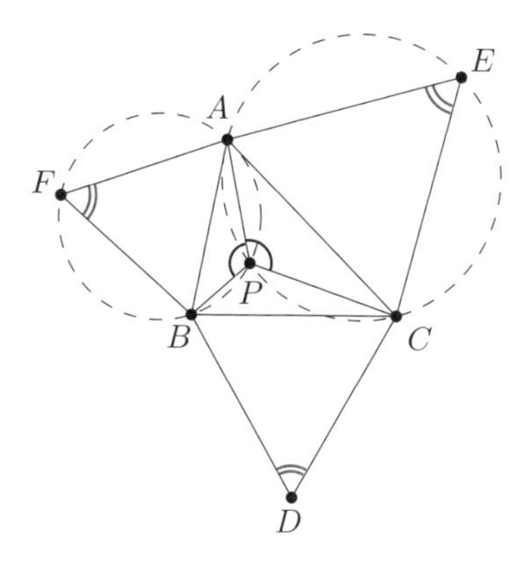

Then $\angle APB = 180° - \angle AFB = 120°$ and likewise $\angle APC = 120°$ hence also $\angle BPC = 120°$ implying that B, D, C, P lie on a single circle. Thus, the three circumcircles indeed pass through a common point.

Next we observe that

$$\angle FPC = \angle FPB + \angle BPC = \angle FAB + \angle BPC = 60° + 120° = 180°.$$

Hence the line CF passes through P and by symmetry BE and AD pass through it too. We may conclude.

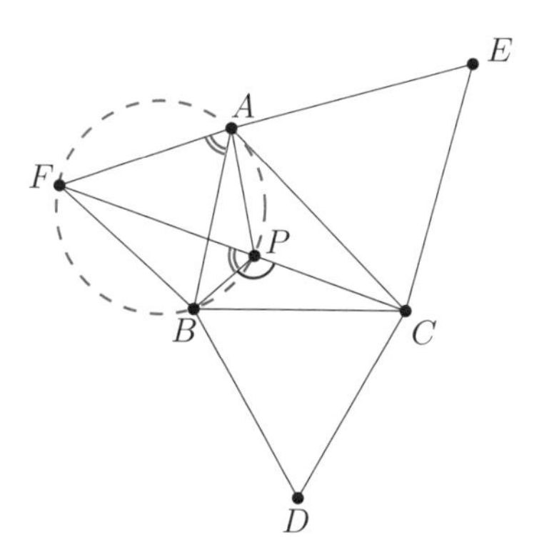

16. [AIME 1989] Triangle ABC has right angle at B, and contains a point P for which $PA = 10$, $PB = 6$, and $\angle APB = \angle BPC = \angle CPA$. Find PC.

Solution. Denote the length of PC by x. Note that since the angles by P are all equal to $120°$, the squares of the side lengths of triangle ABC can be expressed in terms of x by the Law of Cosines applied to triangles ABP, BCP, and CAP.

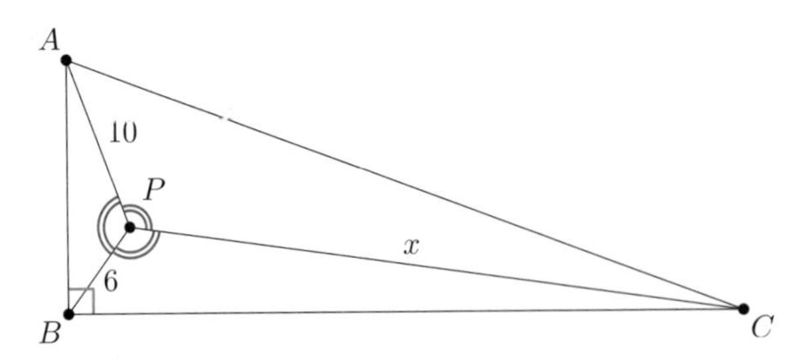

Keeping in mind that $-2\cos 120° = 1$, we obtain

$$AB^2 = 10^2 + 6^2 + 10 \cdot 6,$$
$$BC^2 = 6^2 + x^2 + 6x,$$
$$CA^2 = x^2 + 10^2 + 10x.$$

Finally, using the Pythagorean Theorem in triangle ABC we can form equation in x, which simplifies to $6^2 + 10 \cdot 6 + 6^2 = 4x$, i.e. $x = 33$.

17. [All-Russian Olympiad 2005] Points P, Q are given on the sides AB, AD, respectively, of a parallelogram $ABCD$ ($AB > AD$) such that $AP = AQ = x$. Prove that as x varies, the circumcircles of the triangles PQC pass through another fixed point (other than C).

    Proof. Denote the angle bisector by vertex A by ℓ and let it be vertical. The triangle APQ is isosceles so PQ is horizontal and Q is the reflection of P about ℓ.

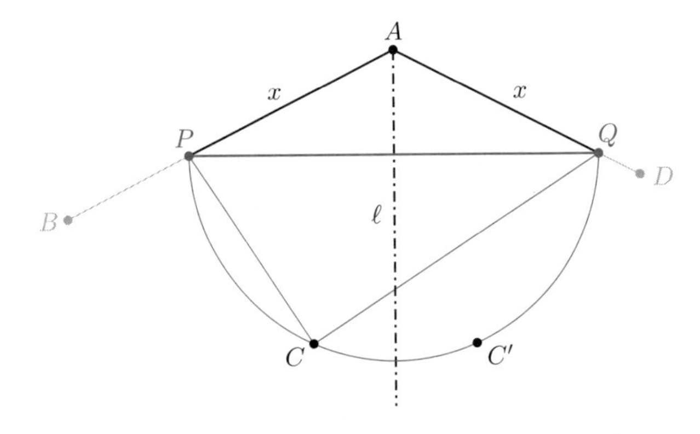

Thus the whole circumcircle of triangle CPQ is symmetric about ℓ and since it passes through C, it also passes through its reflection C' in ℓ, which is the fixed point we were to find (note that $AB > AD$ implies $C' \neq C$).

18. In triangle ABC, medians BB_1 and CC_1 are perpendicular. Given that $AC = 19$ and $AB = 22$, find BC.

First Solution. We draw the third median AA_1 and recall that medians are concurrent at the centroid G. As A_1 is the midpoint of a hypotenuse in right triangle BCG, we have $A_1G = A_1B$ which (since medians "trisect" each other) may be rewritten as $\frac{1}{3}AA_1 = \frac{1}{2}BC$.

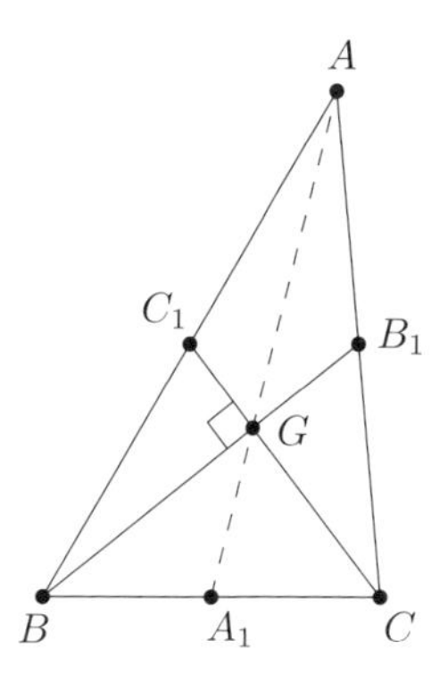

Recalling the median formula (see Corollary 1.24(a)), we square the equality and obtain

$$\frac{1}{9} \cdot \left(\frac{b^2 + c^2}{2} - \frac{a^2}{4} \right) = \frac{1}{4} a^2,$$

which rewrites as $b^2 + c^2 = 5a^2$. Plugging in the numbers, we find $5a^2 = 845$, i.e. $a = 13$.

Second Solution. Again we recall that the centroid G "trisects" medians. Denote their lengths by $BB_1 = 3y$, $CC_1 = 3z$.

Pythagorean theorems in right triangles BGC_1 and CGB_1 yield

$$\left(\frac{c}{2} \right)^2 = 4y^2 + z^2 \quad \text{and} \quad \left(\frac{b}{2} \right)^2 = y^2 + 4z^2.$$

Now we could plug in the values of b and c and solve these equations for y and z but observe that we are only interested in

$$BC^2 = BG^2 + CG^2 = 4y^2 + 4z^2.$$

Hence we sum the equations instead. After multiplying by $\frac{4}{5}$ we get $BC^2 = \frac{1}{5}(b^2 + c^2)$ which gives $BC = 19$ again.

Third Solution. We apply the perpendicularity criterion (see Proposition 1.22) for quadrilateral BCB_1C_1 and obtain

$$a^2 + \left(\frac{a}{2}\right)^2 = \left(\frac{b}{2}\right)^2 + \left(\frac{c}{2}\right)^2,$$

from which we again find $BC = 19$.

19. Let ABC be a right triangle with right angle by C and with $CA = 8$, $CB = 6$. Semicircle with diameter CX where $X \in AC$ touches side AB. Find its radius.

First Solution. Denote the center of the semicircle by O, its radius by r and the point of tangency with AB by D. We express the area of triangle ABC in two different ways.

First, as $\angle ACB = 90°$, the area is simply $\frac{1}{2}AC \cdot BC = 24$. On the other hand, by the Pythagorean Theorem we have $AB = \sqrt{8^2 + 6^2} = 10$ and thus

$$[ABC] = [ABO] + [BCO] = \frac{1}{2}AB \cdot r + \frac{1}{2}BC \cdot r = 8r$$

Equating we get the answer $r = 3$.

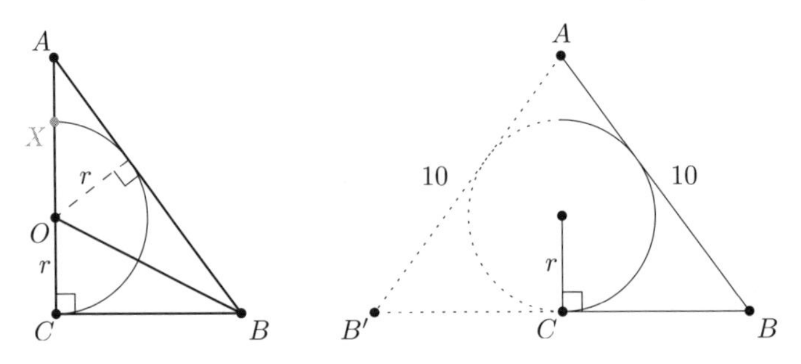

Second Solution. Denote the reflection of B about AC by B'. Then $AB' = AB = 10$, $B'B = 2 \cdot CB = 12$, and r is the inradius of triangle $AB'B$. Recalling the area formulas (see Proposition 1.25), we compute it as

$$r = \frac{[AB'B]}{\frac{1}{2}(AB' + B'B + BA)} = \frac{\frac{1}{2} \cdot 12 \cdot 8}{\frac{1}{2}(10 + 10 + 12)} = 3.$$

20. Four consecutive sides of an equiangular hexagon have lengths 1, 7, 4, and 2. Find the lengths of the remaining two sides.

First Solution. The common value of all the interior angles is $120°$. Hence the four sides can be drawn into a triangular grid made of equilateral triangles of unit side length. The other two sides are then seen to have lengths 6 and 5, respectively.

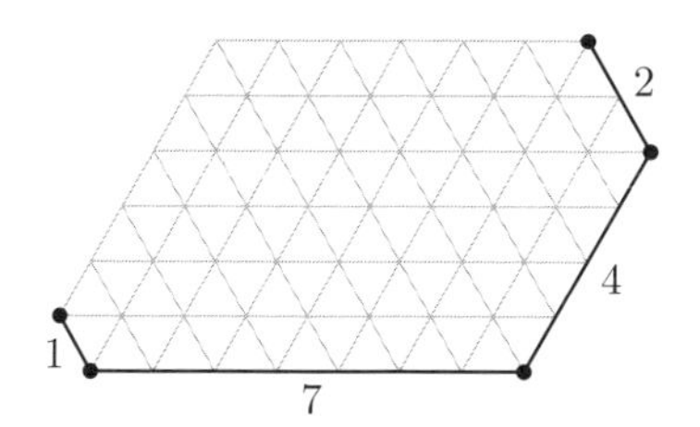

Second Solution. In general, let $ABCDEF$ be a hexagon with side lengths a, b, c, d, e, and f, the first four of which are given, and all interior angles equal to $120°$. Extending its sides AF, BC, DE we form three small equilateral triangles ABX, CDY, EFZ.

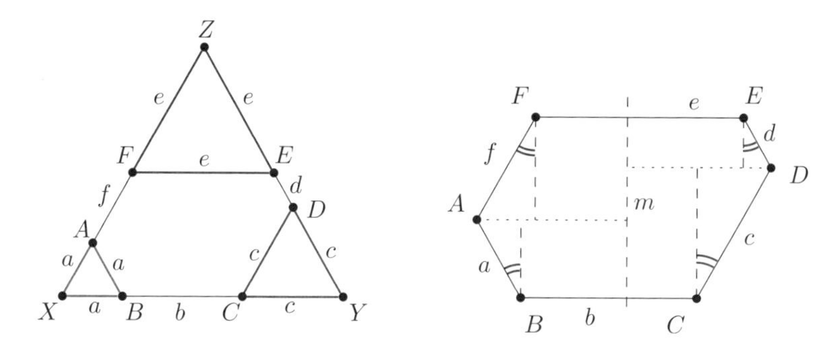

Thus triangle XYZ is equilateral too implying

$$a + b + c = c + d + e = e + f + a.$$

The side lengths e, f are then easily calculated as $e = a + b - d$ and $f = c + d - a$.

Third Solution. Place BC horizontally and observe that the sides AB, CD, DE, and FA all make $30°$ angles with any vertical line m. Hence the lengths of the projections of these sides onto m are proportional to the side lengths. Since B and C project to the same point and so do E and F, we conclude $a + f = c + d$. Hence $f = c + d - a = 5$. Similarly rotating this argument $e = a + b - d = 6$.

21. Let ABC be a triangle with $\angle A = 60°$ and denote its incenter by I. Lines BI, CI intersect the opposite sides at E, F, respectively. Prove that $IE = IF$.

Proof. Recalling that the measure of $\angle BIC$ depends on the measure of $\angle A$ only (see Proposition 1.11) we obtain

$$\angle EIF = \angle BIC = 90° + \frac{1}{2}\angle A = 120°$$

and hence the quadrilateral $AFIE$ is cyclic. As AI is the angle bisector of angle FAE, point I is the midpoint of arc EF of the circumcircle of triangle AEF (see Example 1.8). Thus $IE = IF$ as desired.

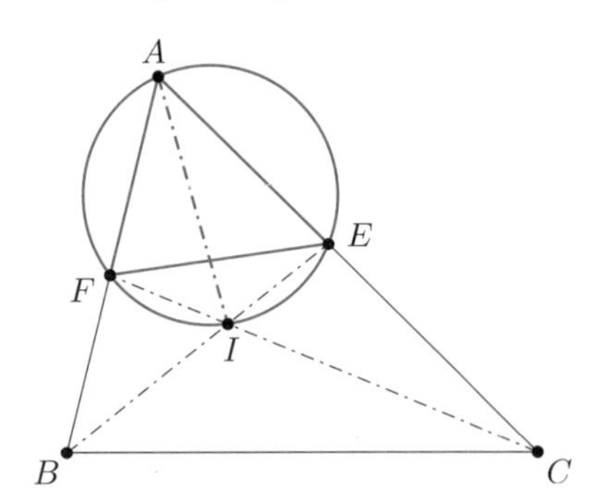

22. [AIME 2005] In quadrilateral $ABCD$ let $BC = 8$, $CD = 12$, $AD = 10$, and $\angle A = \angle B = 60^\circ$. Find the distance AB.

First Solution. Let C_0, D_0 be the feet of perpendiculars dropped to AB from C, D, respectively. Then from right triangle CC_0B, we obtain $C_0B = BC \cdot \cos 60^\circ = \frac{1}{2}BC = 4$ and $CC_0 = BC \cdot \cos 30^\circ = 4\sqrt{3}$. Similarly, we get $AD_0 = \frac{1}{2}AD = 5$ and $DD_0 = 5\sqrt{3}$. Let P be the foot of perpendicular from C to DD_0.

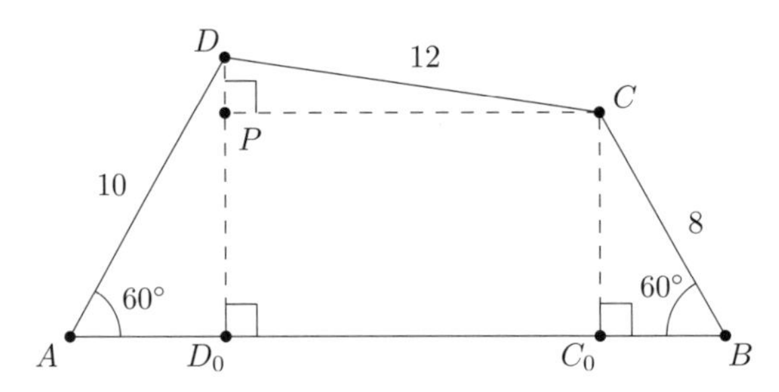

Then $DP = DD_0 - CC_0 = \sqrt{3}$. Rectangle D_0C_0CP gives $D_0C_0 = PC$ and from the Pythagorean Theorem in triangle DPC, we learn $PC = \sqrt{12^2 - (\sqrt{3})^2} = \sqrt{141}$. Hence $AB = AD_0 + D_0C_0 + C_0B = 9 + \sqrt{141}$.

Second Solution. Denote by X the intersection of rays AD and BC. Then the triangle ABX is equilateral. Denote by x its side length.

The Law of Cosines applied to triangle XDC implies

$$\begin{aligned} 12^2 &= (x-8)^2 + (x-10)^2 - 2(x-8)(x-10)\cos 60^\circ \\ 0 &= x^2 - 18x - 60, \end{aligned}$$

which has solutions $9 \pm \sqrt{141}$. Since the length of AX is positive, we have $AB = 9 + \sqrt{141}$.

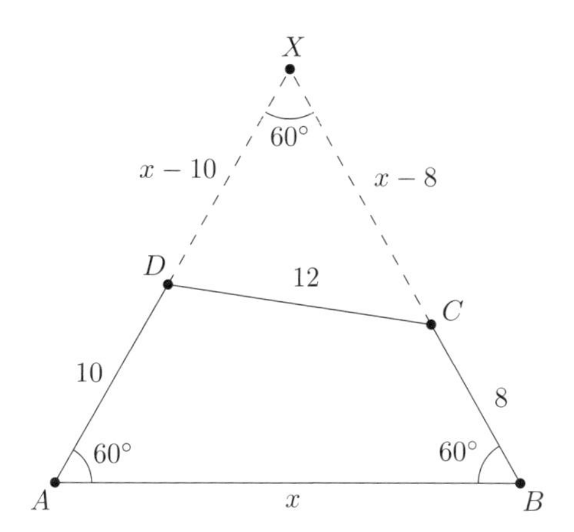

23. Let $ABCD$ be a convex quadrilateral. Find point X for which the sum of distances to its vertices is minimal.

Solution. The point we are looking for is the intersection of diagonals of $ABCD$. Indeed, by triangle inequalities in (possibly degenerate) triangles ACX, BDX we learn

$$AX + XC \geq AC \qquad \text{and} \qquad BX + XD \geq BD.$$

Hence $XA + XB + XC + XD \geq AC + BD$. The equality occurs if it occurs in both partial inequalities, i.e. if X lies on both AC and BD.

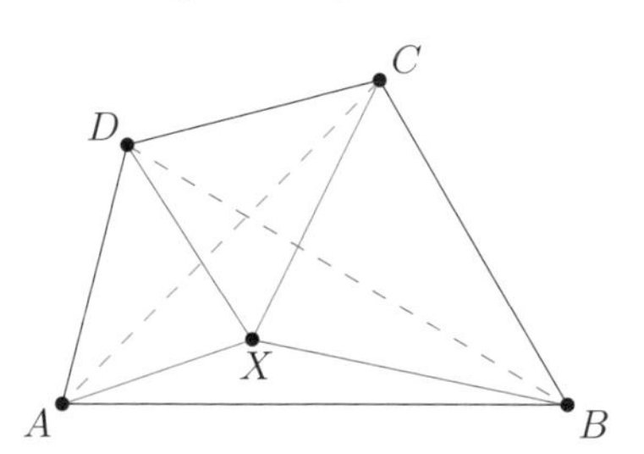

24. Circles ω_1, ω_2 intersect at points A and B. An arbitrary line passing through B intersects ω_1 for the second time at K (outside ω_2) and ω_2 at L (outside ω_1).

   (a) Prove that all possible triangles AKL are similar to each other.

   (b) Let the tangents at points K and L to the respective circles intersect at P. Prove that $KPLA$ is cyclic.

Proof.

(a) Angle $\angle LKA$ is an angle subtending the arc AB on the circle ω_1, hence its magnitude is fixed. Similarly, $\angle ALK$ is fixed and thus all the triangles AKL are similar (AA).

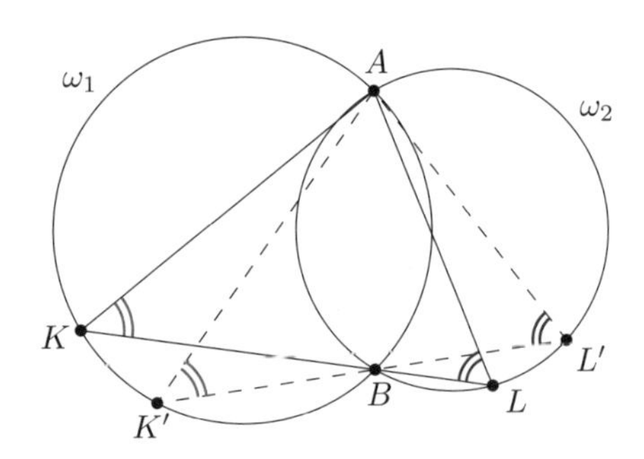

(b) From tangency we obtain, $\angle PKL = \angle KAB$ and $\angle KLP = \angle BAL$ (see Proposition 1.34). Thus,

$$\angle LPK = 180^\circ - \angle PKL - \angle KLP = 180^\circ - \angle KAL$$

and $KPLA$ is cyclic.

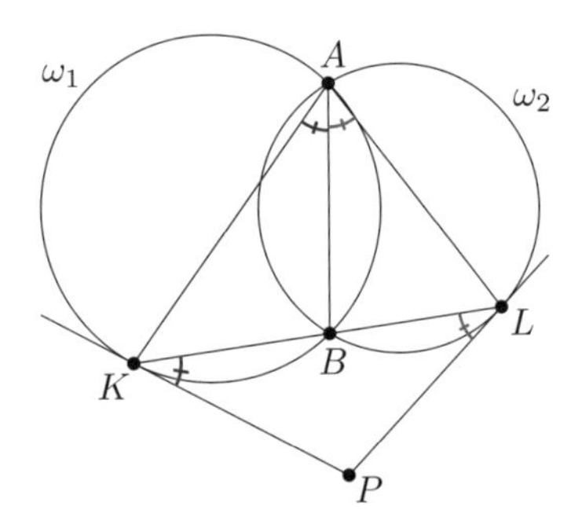

25. Let $ABCD$ be a quadrilateral with $AB \parallel CD$. If $\angle ADB + \angle DBC = 180^\circ$, prove that

$$\frac{AB}{CD} = \frac{AD}{BC}.$$

First Proof. The condition $\angle ADB + \angle DBC = 180^\circ$ suggests employing Law of Sines since $\sin \angle ADB = \sin \angle CBD$.

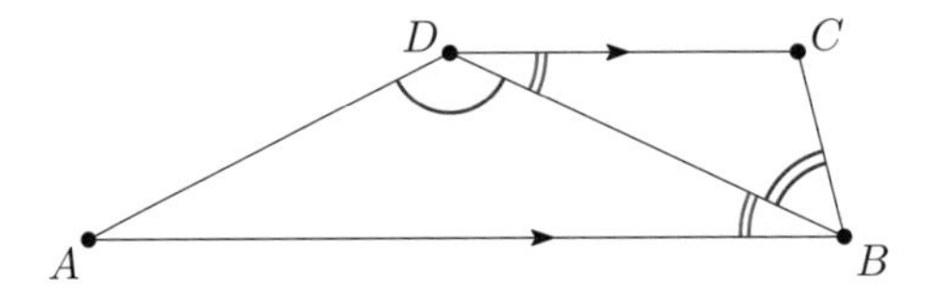

Observing $\angle ABD = \angle BDC$, the Law of Sines in triangles ABD and DBC implies the desired

$$\frac{AB}{AD} = \frac{\sin \angle ADB}{\sin \angle ABD} = \frac{\sin \angle CBD}{\sin \angle BDC} = \frac{CD}{BC}.$$

Second Proof. Let E be the intersection of BC and AD. Then $\angle EDB = 180^\circ - \angle ADB = \angle DBE$ so $ED = EB$.

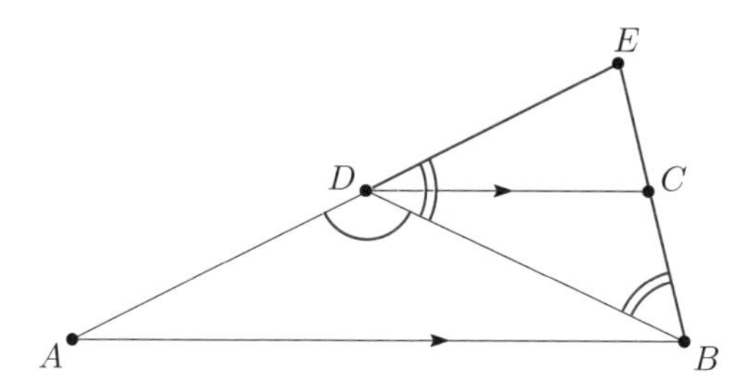

Further, triangles EDC and EAB are similar (AA). Hence

$$\frac{AB}{CD} = \frac{AE}{DE} = \frac{AE}{BE} = \frac{AD}{BC},$$

where in the last equality we again used $AB \parallel CD$.

26. [Sharygin Geometry Olympiad 2012] On side BC of triangle ABC an arbitrary point D is selected. The tangent in D to the circumcircle of triangle ABD meets AC at point B_1. Point C_1 is defined analogously. Prove that $B_1C_1 \parallel BC$.

    Proof. Keeping our diagram nice and clean we choose not to even draw the circumcircles of triangles ABD and ACD. We rewrite the tangency as equality of angles (see Proposition 1.34)

$$\angle CBA = \angle B_1DA \quad \text{and} \quad \angle ACB = \angle ADC_1.$$

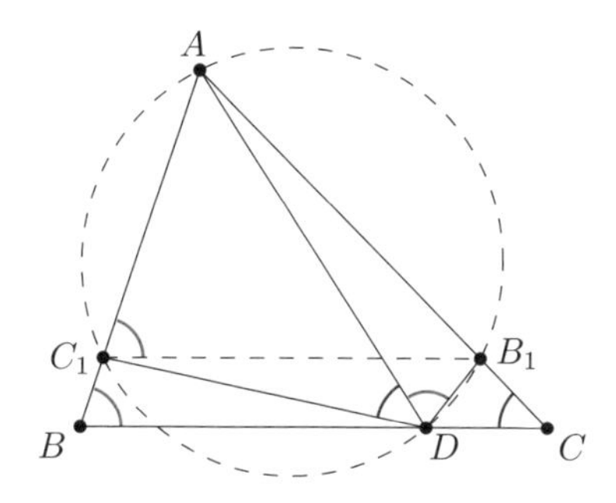

Now we observe that $\angle B_1DC_1 = \angle B + \angle C$, thus it is supplementary to $\angle BAC$, which implies that the quadrilateral AC_1DB_1 is cyclic. Therefore

$$\angle B_1C_1A = \angle B_1DA = \angle B$$

and we indeed have $B_1C_1 \parallel BC$.

27. [J. H. Conway] Conway's[2] circle.

Let ABC be a triangle and denote by A_1, A_2 the points on the rays opposite to AB, AC, respectively, satisfying $AA_1 = AA_2 = BC$. Define points B_1, B_2, C_1, C_2 analogously. Prove that points A_1, A_2, B_1, B_2, C_1, C_2 lie on a single circle.

First Proof. We aim to find point X with the same distance from all the six points.

The locus of points which have the same distance from A_1 and A_2 is the perpendicular bisector of A_1A_2. Since the triangle AA_1A_2 is isosceles, it coincides with the angle bisector of $\angle A$. Thus the only conceivable candidate for X is the incenter I of triangle ABC.

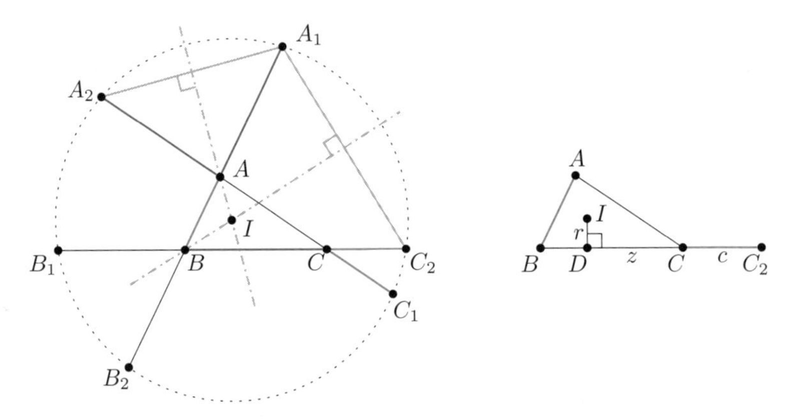

So far we have $IA_1 = IA_2$ and likewise $IB_1 = IB_2$ and $IC_1 = IC_2$. To finish the proof, it suffices to show for instance $IA_1 = IC_2$ (the rest follows by symmetry). But this is the same thing again! As $AA_1 = BC$ and $BA = CC_2$, triangle BA_1C_2 is isosceles and the angle bisector of $\angle B$ and the perpendicular bisector of A_1C_2 coincide. Since I lies on the former, it has the same distance from A_1 and C_2 which completes the proof.

Second Proof. Once we manage to guess that the center of the circle should be the incenter I of triangle ABC, we may observe that its distance to say C_2 is approachable in terms of basic elements of triangle ABC. Indeed, if D is the point of contact of the incircle with the side BC then $DC_2 = DC + CC_2 = z + c = s$ so $IC_2 = \sqrt{r^2 + s^2}$. Since this value is symmetric in a, b, c, we may conclude.

28. Let ABC be an acute triangle. Prove that $h_a > \frac{1}{2}(b + c - a)$, where h_a is the length of A-altitude in triangle ABC.

---

[2]John Horton Conway (1936) is a contemporary British mathematician known for many delightful discoveries both in recreational and research mathematics.

First Proof. Let D be the foot of A-altitude and observe that triangle ABC being acute implies that D lies on the segment BC. Triangle inequalities in the triangles ABD and ACD yield $h_a + BD > c$ and $h_a + DC > b$. Summing them we get $2 \cdot h_a + a > b + c$ and the proof is complete.

Second Proof. Again denote the foot of A-altitude by D and recall that $\frac{1}{2}(b+c-a)$ is the distance from A to the points of contact F, E of the incircle with the sides AB, AC, respectively (see Proposition 1.15).

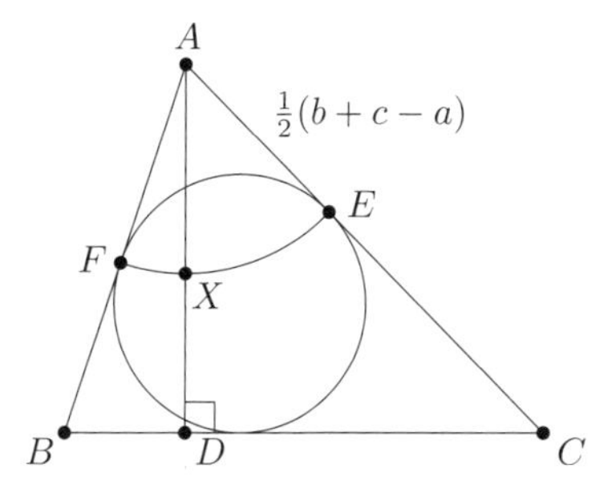

We want to compare the lengths AD and AE. As triangle ABC is acute, the A-altitude intersects the minor arc with center A and endpoints E, F. Denote the intersection by X. Since the whole arc EF lies inside the incircle, point X lies inside it too. Hence it lies inside the triangle ABC and thus on the *segment* AD. Now we are done by

$$\frac{1}{2}(b+c-a) = AE = AX < AD = h_a.$$

Remark. The conclusion is in general not true for obtuse triangles. Find an example!

Remark. Given that triangle ABC is acute, can you prove stronger inequality $h_a > \frac{1}{2}(b+c-a)+r$, where r is the inradius of triangle ABC?

29. Let ABC be a triangle. Find the locus of points X ($X \neq A$) for which the triangles AXB and AXC have equal area.

Solution. We distinguish two cases based on the position of X. If AX intersects the segment (!) BC, we denote by Y the intersection and use the Area Lemma (see Proposition 1.27). We have

$$\frac{[AXB]}{[AXC]} = \frac{YB}{YC},$$

so we need Y to be the midpoint of BC. In other words, in this case the desired points X form a line containing the A-median of triangle ABC.

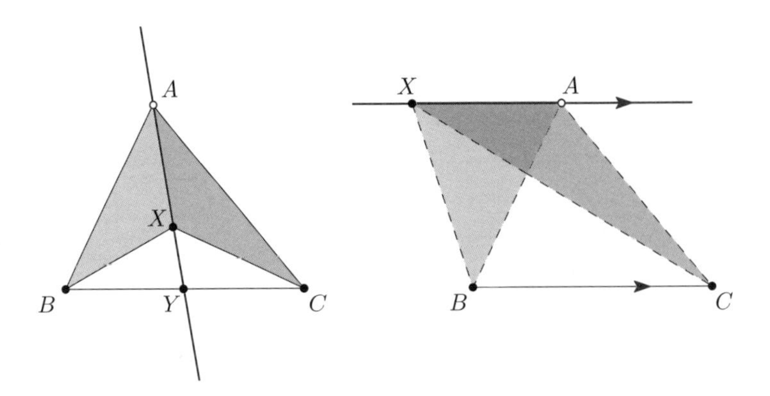

In the other case, points B and C are in the same half-plane with borderline AX. Then since the triangles AXB and AXC have common base AX, we in fact need points B and C to have the same distance from the line AX, which in this case reads as $BC \parallel AX$. Thus, we add the line through A parallel with BC to our locus and the solution is complete.

30. Let $ABCD$ be a quadrilateral with perpendicular diagonals inscribed in a circle with radius R. Prove that

$$AB^2 + BC^2 + CD^2 + DA^2 = 8R^2.$$

Proof. Denote by α, β, γ, δ the inscribed angles corresponding to minor arcs AB, BC, CD, DA, respectively, of the circumcircle of $ABCD$ and rewrite each term on the left hand side by the Extended Law of Sines:

$$(2R\sin\alpha)^2 + (2R\sin\beta)^2 + (2R\sin\gamma)^2 + (2R\sin\delta)^2 = 8R^2 \qquad (: 4R^2)$$
$$\sin^2\alpha + \sin^2\beta + \sin^2\gamma + \sin^2\delta = 2.$$

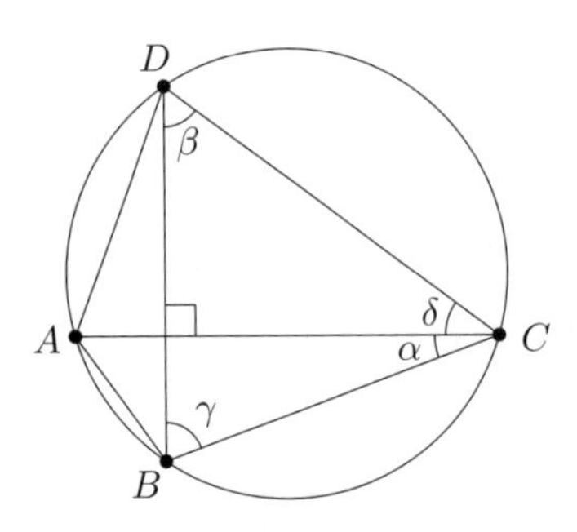

Since the diagonals of $ABCD$ are perpendicular we may rewrite $\sin\delta = \sin(90° - \beta) = \cos\beta$ and $\sin\gamma = \sin(90° - \alpha) = \cos\alpha$. The well-known identity $\sin^2(x) + \cos^2(x) = 1$ then implies the result.

31. [Mexico 1999] A trapezoid $ABCD$ has AB parallel to CD. The external bisectors of $\angle A$ and $\angle D$ meet at P, and the external bisectors of $\angle B$ and $\angle C$ meet at Q. Show that PQ is half the perimeter of $ABCD$.

First Proof. Since P lies on the external angle bisector of $\angle A$, it has equal distance from the lines AB and AD. Similarly, it has equal distance from the lines AD and CD which implies that it lies half the way between the parallel lines AB and CD. Analogous reasoning applies for Q. Thus, if we denote the midpoints of AD, BC by M, N, respectively, then P, M, N and Q are collinear and we can rewrite $PQ = PM + MN + MQ$.

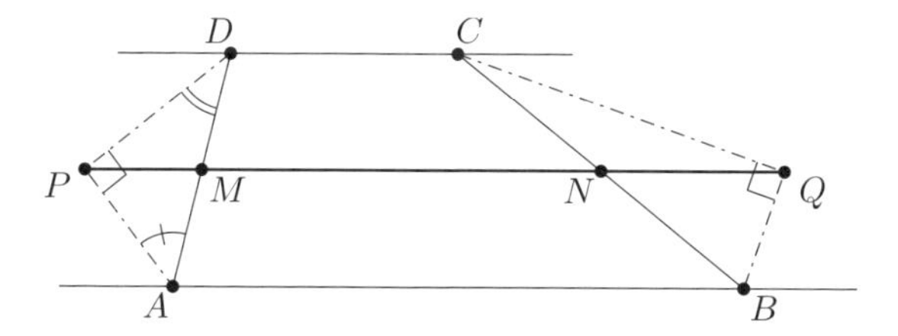

Now observe that as angles by A and D add up to $180°$, the halves of their complements add up to $90°$ which yields $\angle APD = 90°$. Hence M is the circumcenter of right triangle ADP and $MP = \frac{1}{2}AD$. For the same reason $NQ = \frac{1}{2}BC$

Finally, since MN is the midline of trapezoid $ABCD$ we conclude by

$$PQ = PM + MN + NQ = \frac{1}{2}AD + \frac{1}{2}(AB + CD) + \frac{1}{2}BC.$$

Second Proof. Being the intersection of two angle bisectors, P is the center of a circle tangent to lines AB, AD, and CD. Denote by T, U, V the respective points of contact.

Similarly, denote by X, Y, Z the respective points of contact of the lines AB, BC, CD with corresponding circle centered at Q.

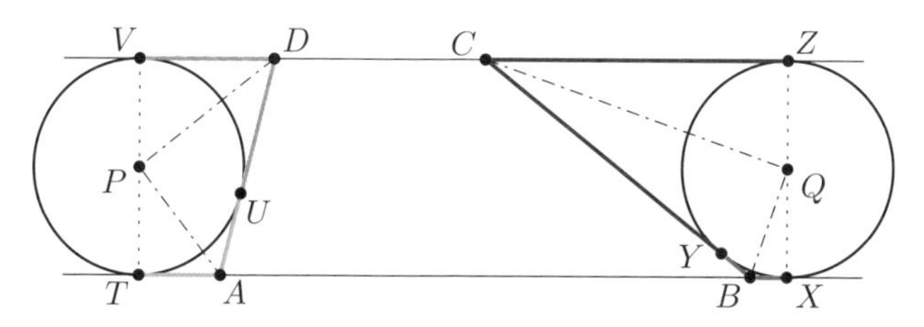

Then P, Q are the midpoints of opposite sides VT, XZ of a rectangle $VTXZ$, respectively, and the result follows by Equal Tangents since

$$\begin{aligned} 2 \cdot PQ = TX + VZ &= (UA + AB + BY) + (UD + DC + CY) \\ &= AB + BC + CD + DA. \end{aligned}$$

32. [AIME 2011] In triangle ABC, $11 \cdot AB = 20 \cdot AC$. The angle bisector of $\angle A$ intersects BC at point D, and point M is the midpoint of AD. Let P be the point of intersection of AC and BM. Find CP/PA.

First Solution. First, we recall the Angle Bisector Theorem and obtain

$$\frac{CD}{DB} = \frac{AC}{AB} = \frac{11}{20}.$$

We may scale up the triangle so that $[BDM] = 20$. Then we can use twice the Area Lemma (see Proposition 1.27) to find the areas $[CMD] = 11$ and $[AMB] = 20$.

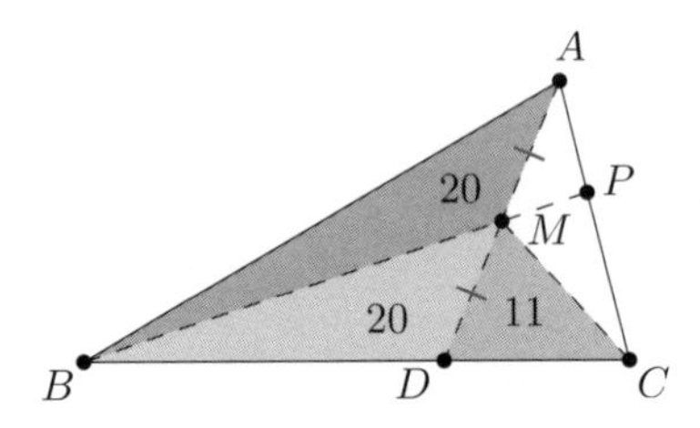

The desired ratio is then found from yet another use of the Area Lemma as

$$\frac{CP}{PA} = \frac{[CMB]}{[AMB]} = \frac{31}{20}.$$

Second Solution. After obtaining $CD/DB = 11/20$ as in the first solution we use Menelaus' Theorem for triangle ADC and line BM to get

$$\frac{AM}{MD} \cdot \frac{DB}{BC} \cdot \frac{CP}{PA} = 1.$$

As $AM/MD = 1$, this can be further rewritten as

$$\frac{CP}{PA} = \frac{BC}{DB} = \frac{CD}{DB} + 1 = \frac{31}{20}.$$

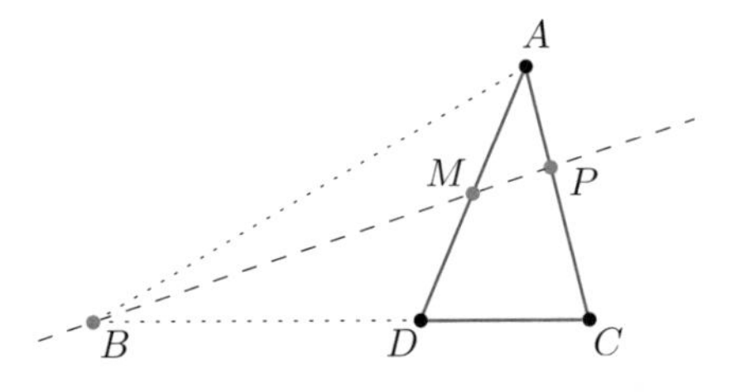

33. A variable segment BC of fixed length d moves such that its endpoints remain on the fixed rays AU, AV. Prove that the circumcircles of all possible triangles ABC are all tangent to a fixed circle.

Proof. The Law of Sines in triangle ABC implies that the circumradius R of triangle ABC is equal to

$$R = \frac{d}{2\sin\angle BAC},$$

so in particular it is the same for all triangles ABC.

Since all these circumcircles pass through A, they are tangent to the circle with radius $2R$ centered at A, which is fixed.

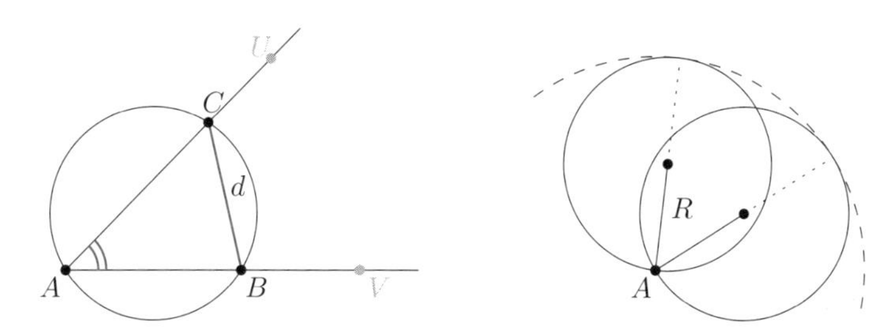

34. Let ABC be a right triangle with $\angle A = 90^\circ$ and altitude AD. Let r, s, t be the inradii of triangles ABC, ADB, and ADC, respectively. Show that $r + s + t = AD$.

First Proof. We divide the desired inequality by AD and note that triangles BAC, BDA, and ADC are pairwise similar (AA). We will use the proportionality of the triangles. The ratio s/AD in triangle BDA corresponds to r/AC in triangle ABC and likewise t/AD in triangle ADC corresponds to r/AB in triangle ABC. We are left to prove

$$\frac{r}{AD} + \frac{r}{AC} + \frac{r}{AB} = 1.$$

but if we denote the incenter of triangle ABC by I, the left-hand side is exactly

$$\frac{[BIC]}{[ABC]} + \frac{[CIA]}{[ABC]} + \frac{[AIB]}{[ABC]},$$

which clearly equals one.

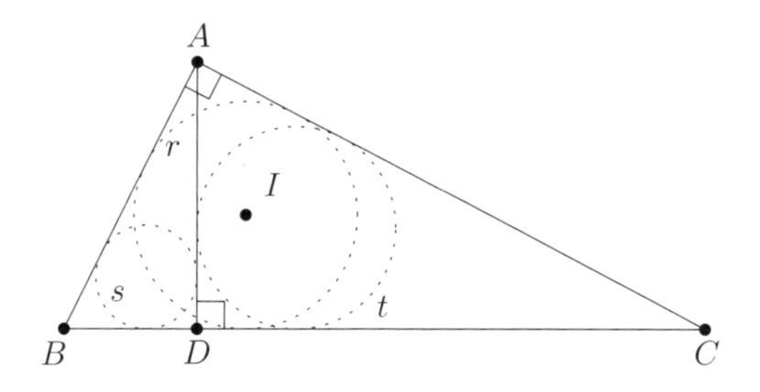

Second Proof. We recall that we have a formula for the inradius of a right triangle (see Proposition 1.16). Applying it three times, we obtain

$$r = \frac{AB + AC - BC}{2}, \; s = \frac{AD + BD - AB}{2}, \; t = \frac{AD + CD - AC}{2}.$$

After addition and some cancelling, we arrive at the result.

35. [AIME 2011] In triangle ABC, $BC = 125$, $CA = 120$, and $AB = 117$. The angle bisector of angle B intersects CA at point K, and the angle bisector of angle C intersects AB at point L. Let M and N be the feet of the perpendiculars from A to CL and BK, respectively. Find MN.

First Solution. Extend AM and AN to meet BC at M' and N', respectively.

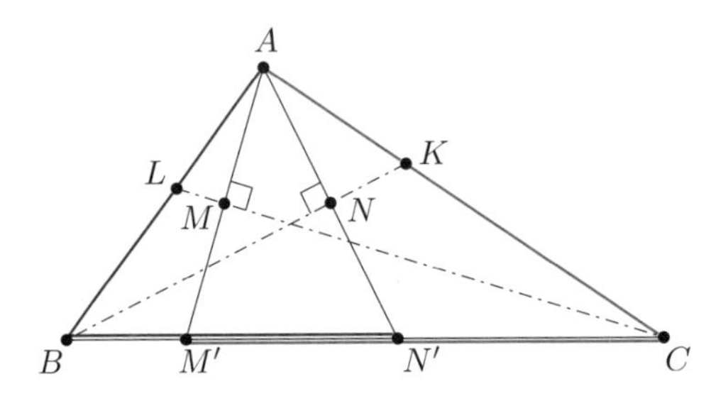

In triangle ACM', the C-altitude and the angle bisector both coincide with CM, hence it is isosceles, $CM' = CA = 120$ and $AM' = 2 \cdot AM$. Analogously, $BN' = AB = 117$ and $AN' = 2 \cdot AN$. Line MN is therefore midline in triangle $AM'N'$ and $M'N' = 2 \cdot MN$. From $M'N' = BN' + M'C - BC$ we infer $MN = \frac{1}{2}(117 + 120 - 125) = 56$.

Second Solution. Observe that $I = BK \cap CL$ is the incenter of triangle ABC and recall the notorious angle $\angle BIC = 90^\circ + \frac{1}{2}\angle A$ (see Proposition 1.11). Also, $MINA$ is cyclic with diameter AI. Then by the Extended Law of Sines in triangle MIN we have

$$MN = AI \sin\left(90^\circ + \frac{1}{2}\angle A\right) = AI \cos\frac{1}{2}\angle A.$$

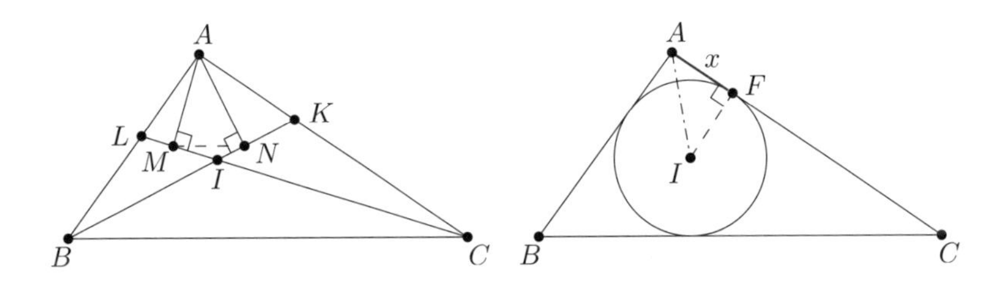

To give some meaning to the expression on the right-hand side we work with a separate diagram. We denote by F the point of contact of the incircle of triangle ABC with AC. Then from right triangle AIF we have $AF = AI \cos \frac{1}{2}\angle A$. Thus $MN = AF = x = \frac{1}{2}(b + c - a) = 56$ (see Proposition 1.15(a) if details are needed).

36. Let $ABCD$ and $AB'C'D'$ be parallelograms such that B' lies on the segment BC and D lies on the segment $C'D'$. Show that their areas are equal.

First Proof. We focus on triangle $AB'D$. It shares the base AD and the corresponding altitude with $ABCD$, thus $[AB'D] = \frac{1}{2}[ABCD]$. But similarly, it has the same base AB' and equal altitude as $AB'C'D'$ so we have also $[AB'D] = \frac{1}{2}[AB'C'D']$. And that's it!

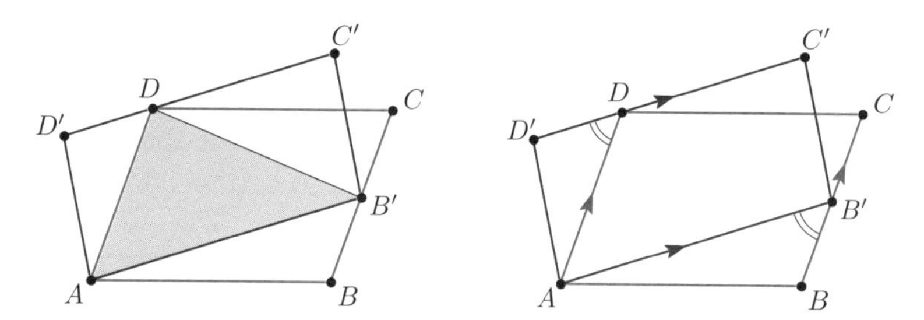

Second Proof. We employ more computational approach. The desired equality $[ABCD] = [AB'C'D']$ rewrites as

$$AB \cdot AD \cdot \sin \angle BAD = AD' \cdot AB' \cdot \sin \angle B'AD'$$

or making use of parallel lines as

$$\frac{AB}{AB'} \cdot \sin \angle B'BA = \frac{AD'}{AD} \cdot \sin \angle AD'D.$$

But from the Law of Sines in triangle ABB' the left-hand side equals just $\sin \angle AB'B$ and likewise from the Law of Sines in triangle $AB'B$, the right-hand side is $\sin \angle D'DA$. Moreover, since $BC \parallel AD$ and $AB' \parallel C'D'$, the two angles are equal and we are done.

37. [St. Petersburg Math Olympiad 1994] In triangle ABC there is a point F on the side AB such that $\angle FAC = \angle FCB$ and $AF = BC$. Further, BE is the internal angle bisector of $\angle B$ with $E \in AC$. Show that $EF \parallel BC$.

Proof. Placing BC horizontally helps to see that we in fact need to prove that points E and F divide the sides AC and AB in the same ratios, since then the result follows by similarity of triangles ABC and

AFE. Thus, taking into account the Angle Bisector Theorem $CE/EA = BC/BA$, it suffices to prove that

$$\frac{BF}{FA} = \frac{BC}{BA}$$

and we may forget the segment BE.

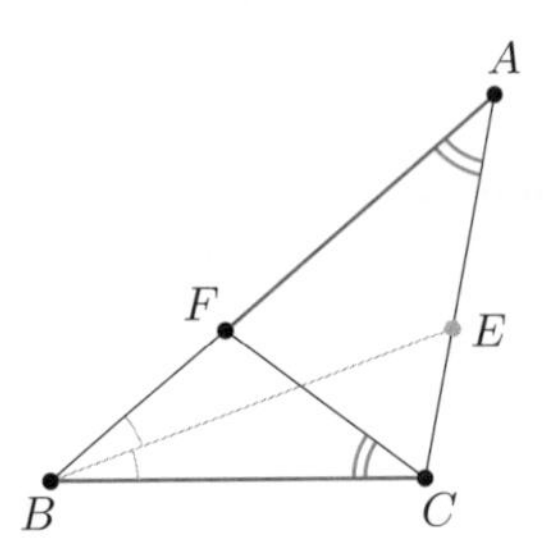

Now using the condition $AF = BC$, this is equivalent to $BF \cdot BA = BC^2$, which is immediate, since $\angle FAC = \angle FCB$ implies that BC is tangent to the circumcircle of triangle AFC (see Proposition 1.34) and the result follows from Power of a Point.

38. Let I be the incenter of triangle ABC. Prove that

$$\frac{AI^2}{bc} + \frac{BI^2}{ca} + \frac{CI^2}{ab} = 1.$$

First Proof. Let D, E, F be the points of contact of the incircle of triangle ABC with the sides BC, CA, AB, respectively.

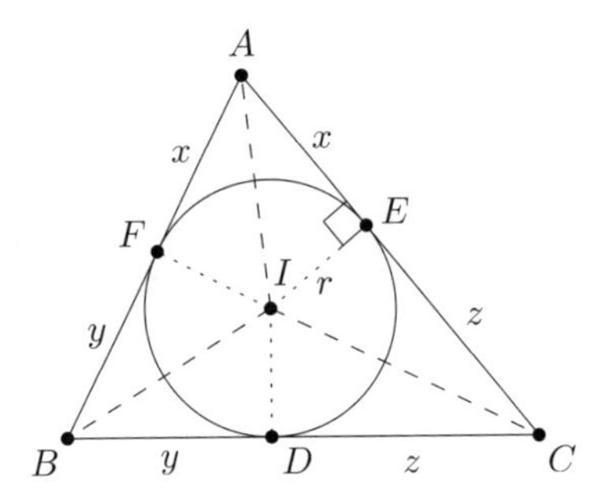

From the right triangle IEA we have $IA^2 = r^2 + x^2$, where the inradius r can be found (see Proposition 1.26) as

$$r^2 = \frac{xyz}{x+y+z}.$$

Now, we can rewrite the desired equality only in terms of x, y, and z, which essentially solves the problem. Indeed,

$$\frac{AI^2}{bc} = \frac{\frac{xyz}{x+y+z} + x^2}{(x+y)(x+z)} = \frac{x(x^2+xy+xz+yz)}{(x+y+z)(x+y)(x+z)} = \frac{x}{x+y+z},$$

and the conclusion follows after we apply the same reasoning for distances BI and CI.

Second Proof. In order to give some meaning to the product in the denominator, we rewrite the first fraction as

$$\frac{AI^2}{bc} = \frac{\frac{1}{2} \cdot AI^2 \sin \angle A}{\frac{1}{2} bc \sin \angle A} = \frac{\frac{1}{2} AI \cdot (AI \sin \angle A)}{[ABC]}.$$

As in the first proof, let D, E, F be the points of contact of the incircle with the sides BC, CA, AB, respectively. Since points E and F lie on a circle with diameter AI, the Extended Law of Sines implies that $AI \sin \angle A = EF$. As E and F are symmetrical about AI, segment EF is perpendicular to AI and thus

$$\frac{\frac{1}{2} AI \cdot (AI \sin \angle A)}{[ABC]} = \frac{\frac{1}{2} AI \cdot EF}{[ABC]} = \frac{[AEIF]}{[ABC]}.$$

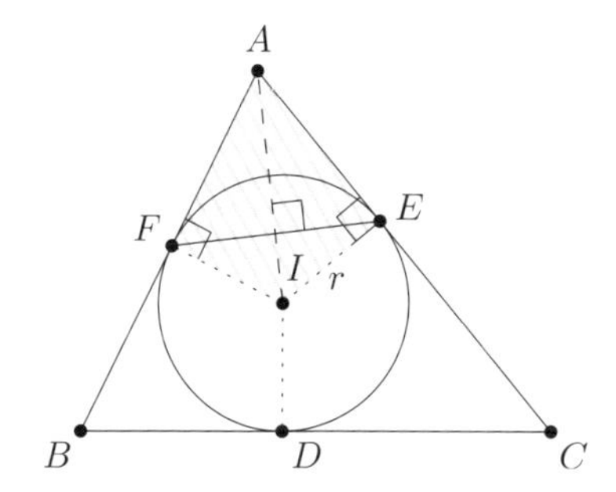

Likewise we obtain

$$\frac{BI^2}{ac} = \frac{[BFID]}{[ABC]} \quad \text{and} \quad \frac{CI^2}{ab} = \frac{[CDIE]}{[ABC]}$$

and the result follows.

Third Proof. We recall that the whole left-hand side can be expressed in terms of the triangle sides. Combining the formula for the length of the angle bisector (see Corollary 1.24) with the known ratio in which it is divided by I (see Corollary 1.28) gives

$$AI^2 = \left(\frac{b+c}{a+b+c}\right)^2 \cdot bc \cdot \left(1 - \left(\frac{a}{b+c}\right)^2\right),$$

from which we after some manipulation find

$$\frac{AI^2}{bc} = \frac{(b+c-a)(a+b+c)}{(a+b+c)^2} = \frac{b+c-a}{a+b+c}.$$

Again, the result follows after applying the same to BI and CI.

Remark. This problem has the following generalization which can be proved using the idea from the second presented proof. If P and Q are isogonal conjugates with respect to triangle ABC and both lie in its interior, then

$$\frac{AP \cdot AQ}{bc} + \frac{BP \cdot BQ}{ca} + \frac{CP \cdot CQ}{ab} = 1.$$

39. In parallelogram $ABCD$ with $\angle BAD > 90^\circ$, show that the circle passing through the projections of C onto AB, BD, and DA, respectively, passes through the center of the parallelogram.

Proof. Denote the feet of projections onto AB, BD, DA by P, Q, R, respectively, and observe that since $\angle CPB = \angle CQB = 90^\circ$, points C, Q, P, B are concyclic and likewise C, Q, R, D are concyclic. For notational purposes, let X be a point on the extension of CQ beyond Q. We angle-chase:

$$\angle PQR = \angle PQX + \angle XQR = \angle CBA + \angle CDA = 2 \cdot \angle CBA.$$

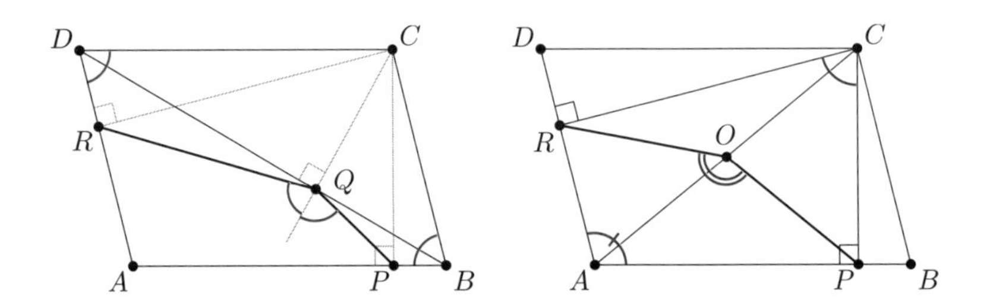

Next, we erase Q and aim to find $\angle POR$ where O is the center of the parallelogram. Since $\angle APC = \angle ARC = 90^\circ$, points A, P, C, R lie on a single circle and O (being the midpoint of its diameter AC) is its center. Hence

$$\angle POR = 2 \cdot \angle PCR = 2 \cdot (180^\circ - \angle DAB) = 2 \cdot \angle CBA$$

and the proof is complete.

40. Let $ABCD$ be a cyclic quadrilateral. Let P be the point on the ray AD such that $AP = BC$ and let Q be the point on the ray AB such that $AQ = CD$. Prove that the line AC cuts PQ at its midpoint.

First Proof. The Law of Sines should be the first idea. Denote the intersection of AC and PQ by X and use the Ratio Lemma (see Proposition 1.18) for triangle PAQ to obtain

$$\frac{PX}{XQ} = \frac{AP}{AQ} \cdot \frac{\sin \angle PAX}{\sin \angle XAQ}.$$

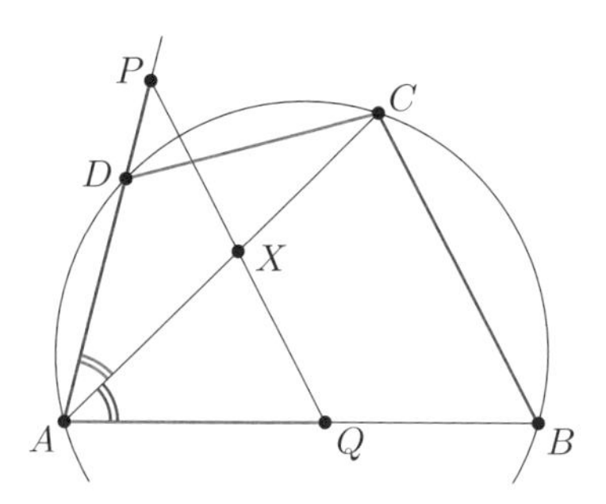

Denoting the radius of the circumcircle of $ABCD$ by R we can use the Extended Law of Sines to learn that $\sin \angle PAX = \sin \angle DAC = CD/2R = AQ/2R$ and likewise $\sin \angle XAQ = AP/2R$. Plugging it in, the whole right-hand side reduces to 1. Hence $PX = XQ$ as desired.

Second Proof. (by Richard Stong) Let P' be the other point on line AD with $AP' = AP = BC$.

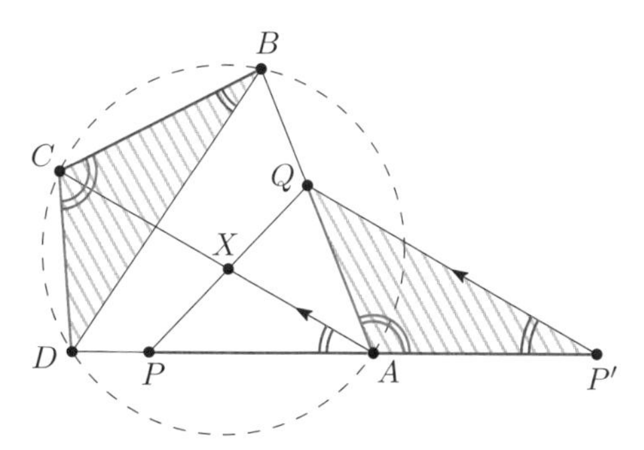

Then as $ABCD$ is cyclic, we have $\angle QAP' = \angle BCD$ and triangles QAP' and DCB are congruent (SAS). Hence $\angle QP'A = \angle DBC = \angle DAC$ and therefore $P'Q$ is parallel to AC. Now in triangle $\Delta PP'Q$, A is the midpoint of PP' and AC is parallel to $P'Q$. Hence AC is the midline and in particular $X = AC \cap PQ$ is the midpoint of PQ.

41. [Junior Balkan 2009] Let $ABCDE$ be a convex pentagon such that $AB + CD = BC + DE$ and a circle ω with center O on the side AE is tangent to the sides AB, BC, CD and DE at points P, Q, R and S, respectively. Prove that the lines PS and AE are parallel.

Proof. First, we use Equal Tangents to transform the given metric condition into something more approachable.

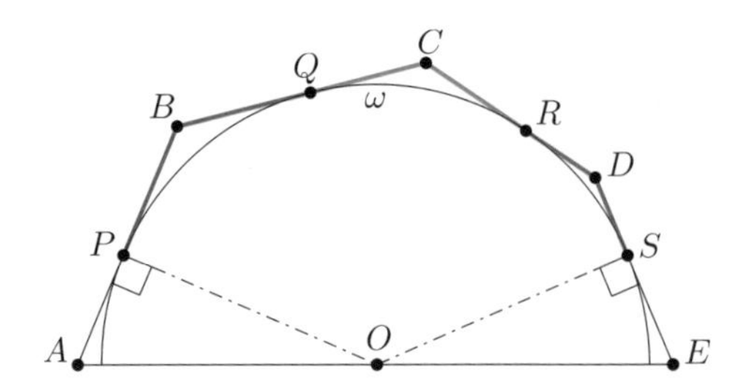

We have

$$AB + CD = AP + PB + CR + RD,$$
$$BC + DE = BQ + QC + DS + SE = PB + CR + RD + SE,$$

thus by comparison, we obtain $AP = SE$. Now, the right triangles AOP and EOS are congruent (SAS) and so they have equal corresponding altitudes. Then the points P and S have the same distance from the line AE (and lie in the same half-plane) implying the desired $PS \parallel AE$.

42. Let P be a point inside acute-angled triangle ABC with $\angle BPC = 180 - \angle A$. Denote by A_1, B_1, C_1 its reflections over the sides BC, CA, AB, respectively. Prove that the points A, A_1, B_1, C_1 are concyclic.

Proof. We angle-chase.

First, observe that

$$\angle C_1AB_1 = \angle C_1AP + \angle PAB_1 = 2 \cdot (\angle BAP + \angle PAC) = 2\angle A$$

and in the same vein we prove that

$$\angle A_1BC_1 = 2\angle B, \qquad \angle B_1CA_1 = 2\angle C.$$

Moreover, we note that $BA_1 = BC_1$ as both distances are by symmetry equal to BP and similar reasoning shows $CA_1 = CB_1$. Keeping in mind we only need to find angle $B_1A_1C_1$ and that we know $\angle CA_1B = \angle BPC = 180° - \angle A$, we lose no information if we erase points P and A.

Now we clearly see the isosceles triangles A_1BC_1 and $\angle B_1CA_1$, which give us

$$\angle C_1A_1B = 90° - \angle B \quad \text{and} \quad \angle CA_1B_1 = 90° - \angle C$$

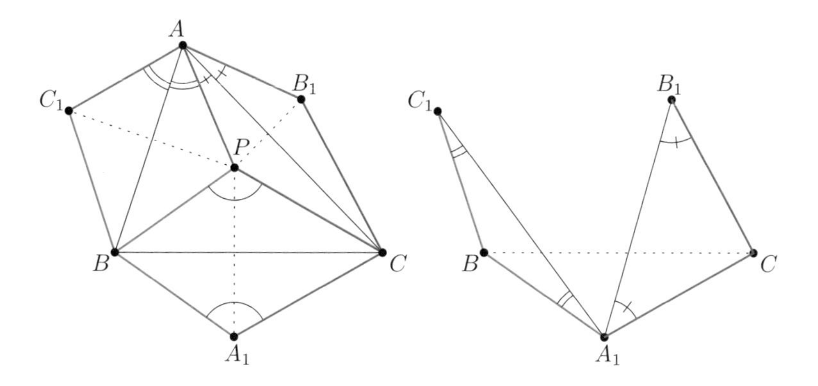

and we can comfortably cross the finish line:

$$\angle B_1A_1C_1 = (180^\circ - \angle A) - (90^\circ - \angle B) - (90^\circ - \angle C) = 180^\circ - 2\angle A.$$

Then points A, A_1, B_1, C_1 are indeed concyclic.

43. Triangle KLM lies inside triangle ABC so that points K, L, M lie on the segments CL, AM, BK, respectively. Prove that the circumcircles of the triangles ABM, BCK, CAL pass through a common point.

First Proof. Let X be the second intersection of the circumcircles of triangles ABM and BCK. We will focus on external angles in triangle KLM, whose sum is 360°.

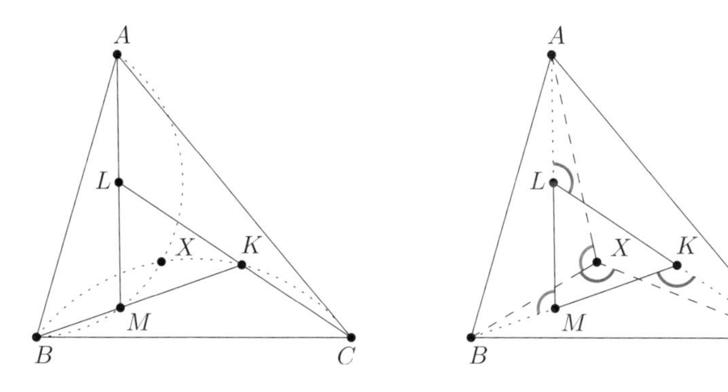

Since $ABMX$ and $BCKX$ are cyclic, we have

$$\angle BKC = \angle BXC, \quad \angle AMB = \angle AXB,$$

so we can find angle CXA as

$$\angle CXA = 360^\circ - \angle BXC - \angle AXB = 360^\circ - \angle BKC - \angle AMB = \\ = \angle CLA.$$

Thus $CALX$ is cyclic and we are done.

Second Proof. Again, let X be the second intersection of the circumcircles of triangles ABM and BCK. Using the cyclic quadrilaterals $ABMX$ and $BCKX$ we can conclude with the following angle chase

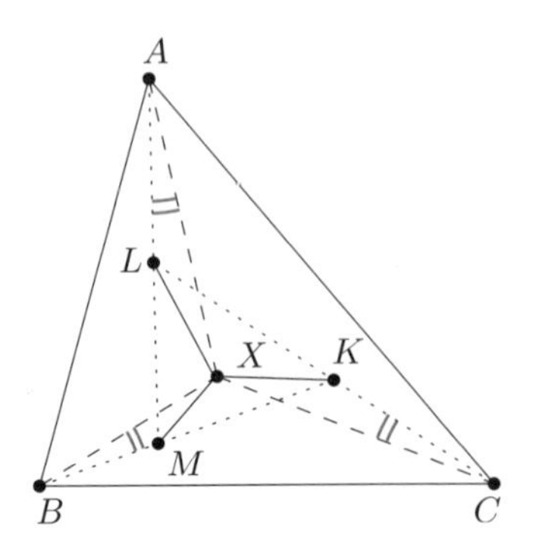

$$\angle LCX \equiv \angle KCX = \angle KBX \equiv \angle MBX = \angle MAX \equiv \angle LAX.$$

44. Let the pentagon $ABCDE$ inscribed in circle ω satisfy $BA = BC$. The line joining $P = BE \cap AD$ and $Q = CE \cap BD$ intersects ω at points X, Y. Prove that $BX = BY$.

Proof. We may assume B is the "bottom" point of ω. From $BA = BC$ we deduce that line AC is horizontal. We are to show that $BX = BY$, i.e. that XY is horizontal too. Thus it suffices to prove $PQ \parallel AC$. We have just gotten rid of the points X and Y.

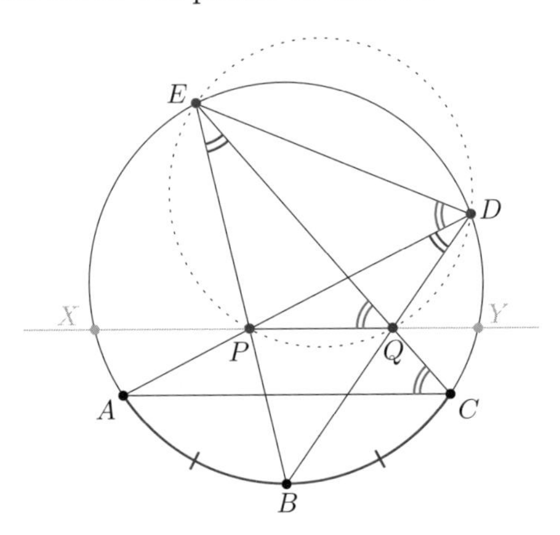

Since angles ADB and BEC subtend equal arcs, they are equal themselves. Thus, $PQDE$ is cyclic. Focusing on angles by the line CE we obtain the desired

$$\angle EQP = \angle EDP \equiv \angle EDA = \angle ECA.$$

45. [Poland 2010] In the convex pentagon $ABCDE$ all interior angles have the same measure. Prove that the perpendicular bisector of segment EA, the perpendicular bisector of segment BC and the angle bisector of $\angle CDE$ intersect at one point.

Proof. The trick is to extend AB to meet CD, DE at X, Y, respectively. The triangles BCX and AEY are then isosceles so the perpendicular bisector of EA is the same line as the angle bisector of angle DYX and likewise, the perpendicular bisector of BC coincides with the angle bisector of angle YXD. The three lines thus concur at the incenter of triangle YXD.

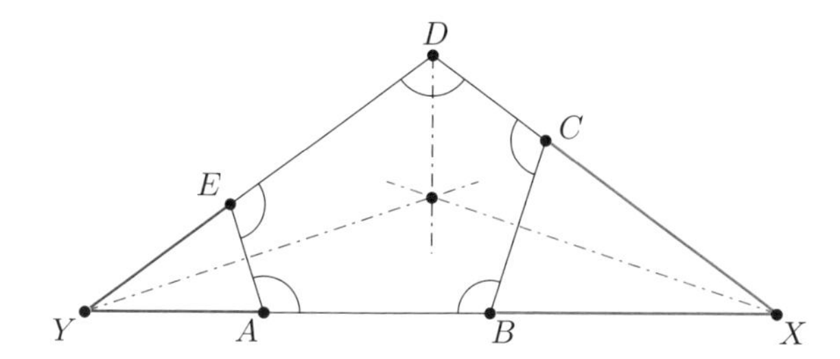

46. Let ω_1, ω_2 be two circles. One of their common external tangents is tangent to ω_1 at A, the second one is tangent to ω_2 at D. Line AD intersects the circles ω_1, ω_2 for the second time at B, C, respectively. Prove that $AB = CD$.

Denote the points of contact of the first tangent with ω_2 by F and of the second tangent with ω_1 by E. We offer two approaches.

First Proof. By symmetry, $ED = AF$ and Power of a Point instantly gives

$$AC \cdot AD = p(A, \omega_2) = AF^2 = ED^2 = p(D, \omega_1) = BD \cdot AD.$$

Cancelling AD we get $AC = BD$. Hence $AB = CD$.

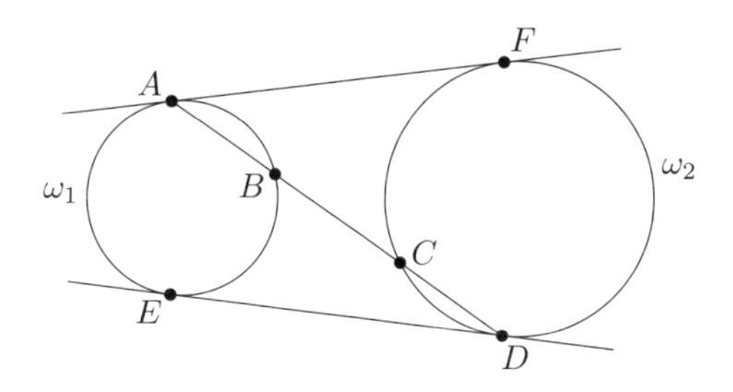

Second Proof. Let P be the midpoint of AD. If we prove that P lies on the radical axis of ω_1, ω_2, we are done, since from $PA \cdot PB = PC \cdot PD$

and $PA = PD$ we easily conclude $PB = PC$ and $AB = CD$. Since the midpoints M, N of AF, DE, respectively, lie on the radical axis (indeed, $MA^2 = MF^2$ and $NE^2 = ND^2$), it is enough to prove M, N, P collinear. But that's immediate as all three points lie on the midline of trapezoid $AEDF$.

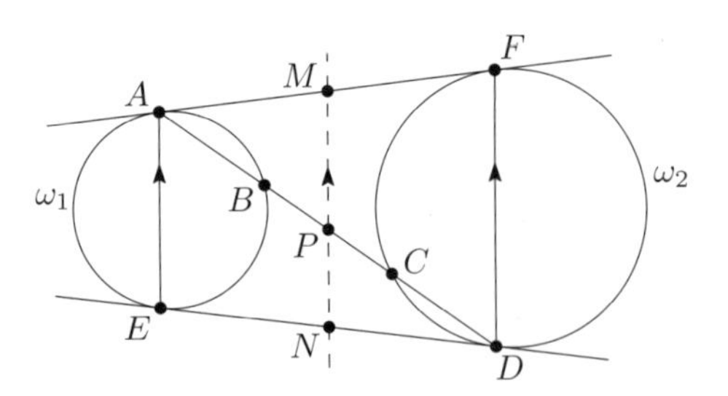

47. [AMC12 2011] Triangle ABC has $AB = 13$, $BC = 14$, and $CA = 15$. The points D, E, and F are the midpoints of BC, CA, and AB, respectively. Let $X \neq D$ be the intersection of the circumcircles of triangles BDF and CDE. What is $XA + XB + XC$?

Solution. We will prove that X is in fact the circumcenter of triangle ABC. First, we note that triangles BDF and DCE are congruent (SSS) and thus their circumcircles have equal radii.

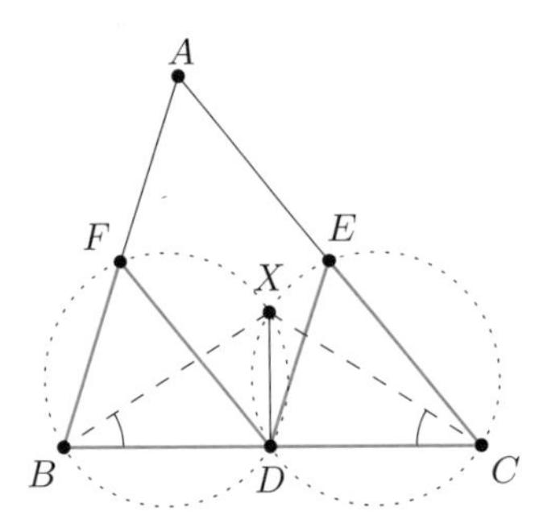

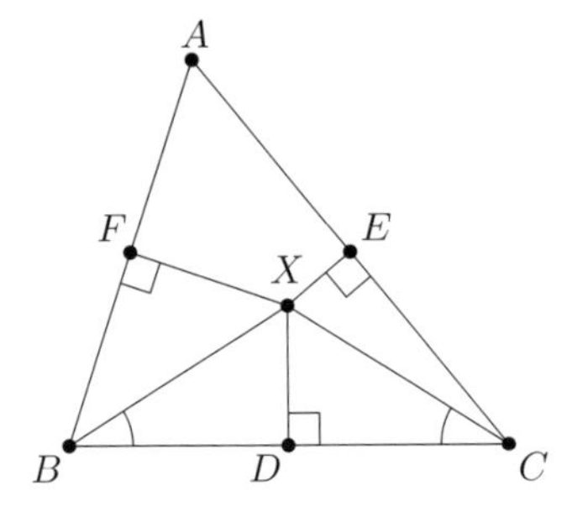

Hence the inscribed angles which correspond to the common chord XD are the same in both circles. Thus $\angle DBX = \angle XCD$, implying that triangle BCX is isosceles. Moreover, XD as a median in an isosceles triangle is perpendicular to the base BC. Thus, XB and XC are diameters of the respective circles, which also implies $\angle BFX = 90°$ and $\angle XEC = 90°$. But then DX, EX, and FX are perpendicular bisectors in triangle ABC, so their intersection X is indeed the circumcenter of triangle ABC.

Denoting the circumradius of triangle ABC by R, we are asked to find

$3R$. We recall the xyz formula for R (see Proposition 1.26)

$$R = \frac{(x+y)(y+z)(z+x)}{4\sqrt{xyz(x+y+z)}},$$

and plug in $x = 7$, $y = 6$, and $z = 8$. The result is $3R = 195/8$.

48. Let $ABCD$ be a quadrilateral with segments BC and AD equal and AB not parallel to CD. Denote by M, N the midpoints of BC and AD, respectively. Prove that the perpendicular bisectors of AB, MN, and CD pass through a common point.

    Proof. Two important ideas lie behind the following short solution. First, the perpendicular bisector is just a locus of equidistant points, and second, if we are proving concurrence of three lines, we often start by intersecting two of them.

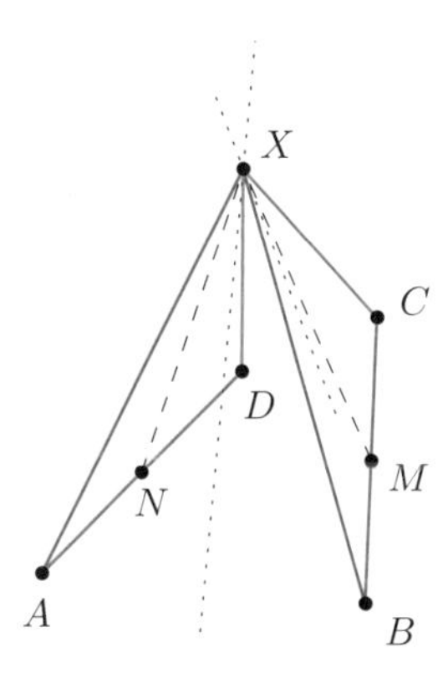

Let X be the intersection of the perpendicular bisectors of AB and CD. Moreover, it suffices to keep in mind just $XA = XB$ and $XC = XD$ and not even bother drawing the perpendicular bisectors. But now triangles XBC and XAD are congruent (SSS)! It follows that the corresponding medians XM and XN are also equal, or in other words, that X lies on the perpendicular bisector of MN.

49. Carnot's[3] Theorem.

    Let X, Y, and Z lie on the sides BC, CA, AB, respectively, of a triangle ABC. Show that the perpendiculars from X, Y, Z to the respective triangle sides meet at one point if and only if

$$BX^2 + CY^2 + AZ^2 = CX^2 + AY^2 + BZ^2.$$

[3]Lazare Nicolas Marguerite Carnot (1753–1823) was an amateur mathematician and French minister of war during the French revolutionary wars.

Proof. First, assume the perpendiculars are concurrent at P. From perpendicularity criterion (see Proposition 1.22) for $PX \perp BC$, we learn

$$BP^2 - PC^2 = BX^2 - CX^2$$

and analogously we find

$$CP^2 - PA^2 = CY^2 - AY^2 \quad \text{and} \quad AP^2 - PB^2 = AZ^2 - BZ^2.$$

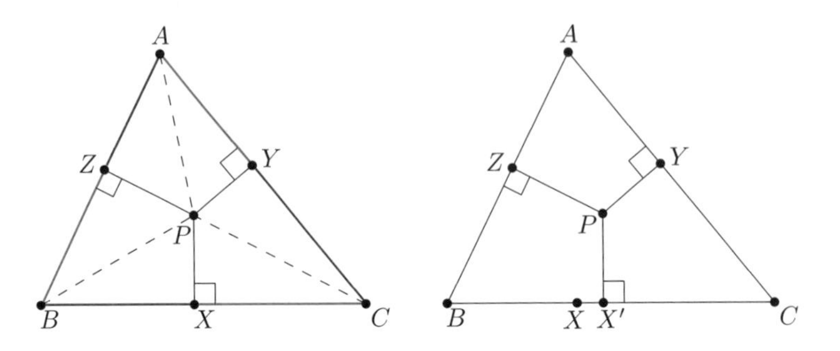

Now if we add the three relations, we obtain the desired

$$BX^2 + CY^2 + AZ^2 = CX^2 + AY^2 + BZ^2.$$

Conversely, assume the metric relation holds for some X, Y, and Z on the triangle sides. Let P be the intersection of the perpendiculars from Y to AC and from Z to AB and let X' be the projection of P to BC. Now for points X', Y, and Z we may use the first part of this statement and obtain

$$BX'^2 + CY^2 + AZ^2 = CX'^2 + AY^2 + BZ^2,$$

which after comparison with the given condition yields

$$BX'^2 - CX'^2 = BX^2 - CX^2.$$

We claim this can only happen if $X = X'$. Indeed, should we have $BX < BX'$, then $CX > CX'$ and the left-hand side has greater value. The case $BX > BX'$ is treated in the same fashion. We may now conclude.

50. [South Africa 2003] In a given pentagon $ABCDE$, triangles ABC, BCD, CDE, DEA and EAB all have the same area. The lines AC and AD intersect BE at points M and N. Prove that $BM = EN$.

    Proof. The equality $[BCD] = [CDE]$ means that as the triangles have common base CD, they also have the same altitude. In other words

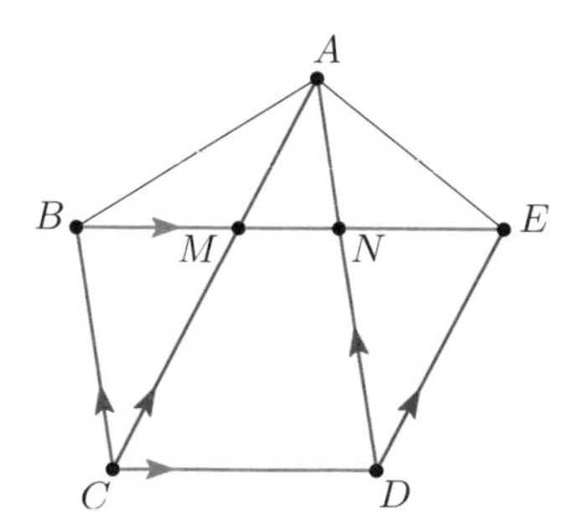

$BE \parallel CD$. Similarly, we find that $CA \parallel DE$ and $BC \parallel AD$ implying that the triangles CMB and DEN are similar (AA).

Moreover, as $BE \parallel CD$, they have the same altitude from C and D, respectively, hence they are in fact congruent. The conclusion follows.

51. Let ABC be a non-right triangle with orthocenter H and let M, N be points on its sides AB and AC. Prove that the common chord of circles with diameters CM and BN passes through H.

Proof. We observe that the circle with diameter CM passes through the foot C_0 of altitude from C and similarly the circle with diameter BN passes through B_0, the foot of altitude from B.

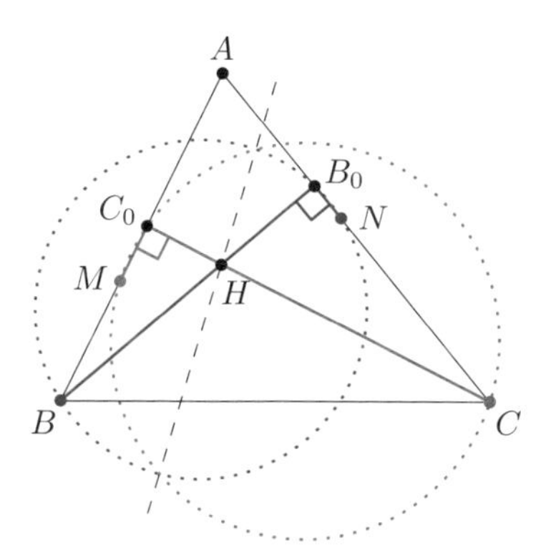

Then since quadrilateral BCB_0C_0 is cyclic (with diameter BC), the Radical Lemma (see Proposition 1.43) implies H indeed lies on the radical axis of the two circles.

52. Let fixed points A, Z, B lie on a line ℓ in this order such that $ZA \neq ZB$. A variable point $X \notin \ell$ and a variable point Y on the segment XZ are chosen. Let $D = BY \cap AX$ and $E = AY \cap BX$. Prove that all lines DE pass through a fixed point.

Proof. We will prove that all the lines DE meet the line AB at the same point. Indeed, let's denote by T the intersection of these lines (which are not parallel, as $ZA \neq ZB$, if in doubt see Example 1.20) and aim to compute the ratio AT/TB.

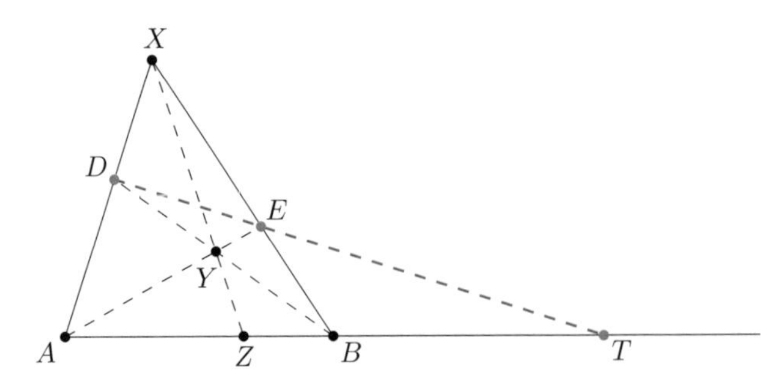

The correct technique here is to compare Ceva's Theorem for cevians passing through Y with Menelaus' Theorem for the collinear points D, E, T, both with respect to triangle ABX. We obtain

$$\frac{AZ}{ZB} \cdot \frac{BE}{EX} \cdot \frac{XD}{DA} = 1 \quad \text{and} \quad \frac{AT}{TB} \cdot \frac{BE}{EX} \cdot \frac{XD}{DA} = 1.$$

Comparing the two gives

$$\frac{AZ}{ZB} = \frac{AT}{TB},$$

which shows that the ratio AT/TB is independent of the positions of X and Y. Moreover, since T lies outside the segment AB, this ratio determines it uniquely. Hence all the lines DE pass through T.

53. Let ω_1 and ω_2 be two circles centered at distinct points O_1 and O_2 and with radii r_1, r_2, respectively.

    (a) Find the locus of points X for which $p(X, \omega_1) - p(X, \omega_2)$ is constant.

    (b) Find the locus of points X for which $p(X, \omega_1) + p(X, \omega_2)$ is constant.

Solution.

(a) Assume for some two points X and Y, we have

$$p(X, \omega_1) - p(X, \omega_2) = p(Y, \omega_1) - p(Y, \omega_2).$$

From the very definition of Power of a Point we obtain $p(X, \omega_1) = O_1X^2 - r_1^2$ and similarly we rewrite the remaining three terms. After some cancellation we get

$$O_1X^2 - O_2X^2 = O_1Y^2 - O_2Y^2,$$

which means (by Proposition 1.22) that $XY \perp O_1O_2$. In other words, we see that the desired points must lie on a line perpendicular to O_1O_2. By reversing the previous calculation, we see that all such points satisfy the condition.

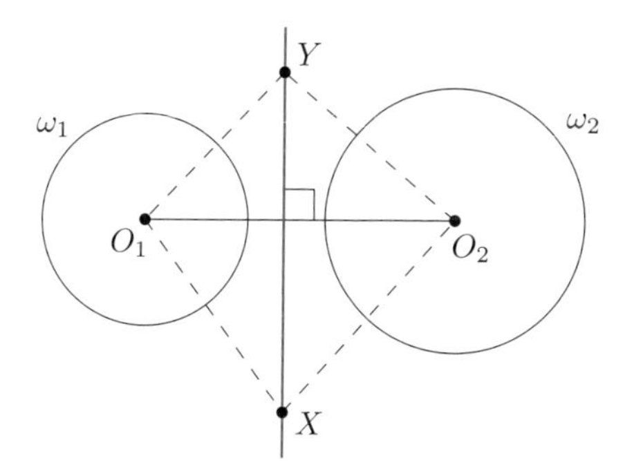

(b) Again we make use of the definition of Power of a Point. After writing $p(X, \omega_1) + p(X, \omega_2) = XO_1^2 - r_1^2 + XO_2^2 - r_2^2$ we see that in fact we need $XO_1^2 + XO_2^2$ to be constant. Now the trick is to look at the midpoint M of O_1O_2. By the median formula (see Proposition 1.24) we have

$$XM^2 = \frac{1}{2}(XO_1^2 + XO_2^2) - \frac{1}{4}O_1O_2^2,$$

which means that the desired locus is formed by points with constant distance XM. In other words, it is a circle centered at M. The case when the triangle XO_1O_2 degenerates into a line is treated in the same vein and the resulting points complete the circle.

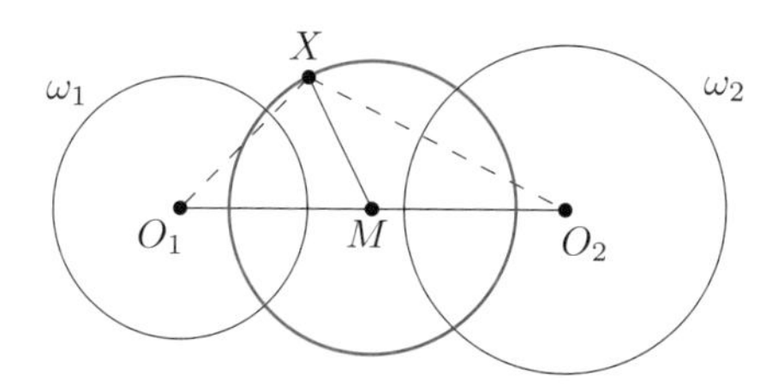

Chapter 5

Solutions to Advanced Problems

1. [Romania 2004] On the sides AB and AD of the rhombus $ABCD$ consider the points E and F such that $AE = DF$. Let $BC \cap DE = P$ and $CD \cap BF = Q$. Prove that points P, A, and Q are collinear.

   Proof. Pairs of parallel lines and equal segments suggest approaching the problem in terms of ratios.

   Observe $PB \parallel AD$ and $BA \parallel DQ$. If we prove that the triangles PBA and ADQ are similar, then the corresponding sides PA and AQ are also parallel and hence P, A, Q are collinear. To that end it suffices to prove $PB/AD = BA/DQ$ (SAS).

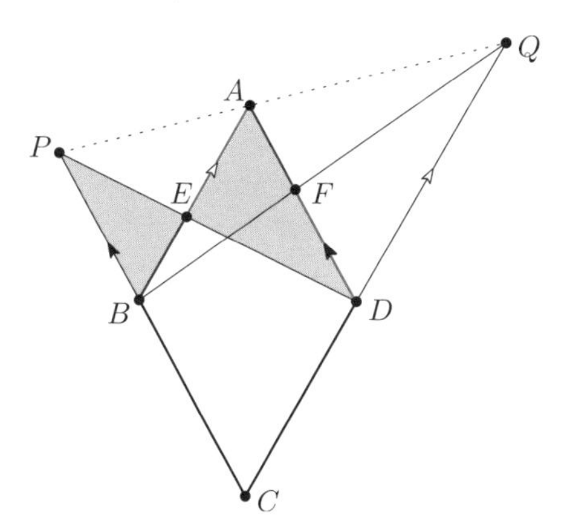

This is easily accomplished once we realize that $\triangle EPB \sim \triangle EDA$ (AA) and likewise $\triangle FQD \sim \triangle FBA$, since then

$$\frac{PB}{AD} = \frac{BE}{EA} = \frac{AF}{FD} = \frac{BA}{DQ}.$$

2. [Switzerland 2011] Let $ABCD$ be a parallelogram such that the triangle ABD is acute and has orthocenter H. The line through H parallel to AB cuts AD and BC at Q and P, respectively, while the line through H parallel to BC cuts AB and CD at R and S, respectively. Prove that the points P, Q, R, S lie on the same circle.

   Proof. We choose to define the orthocenter as the intersection of altitudes DD_0 and BB_0, also we exclude the diagonal BD from our diagram.

   We aim to prove concyclicity in the language of ratios. Namely, by proving
   $$HQ \cdot HP = HS \cdot HR,$$
   which suffices by Power of a Point.

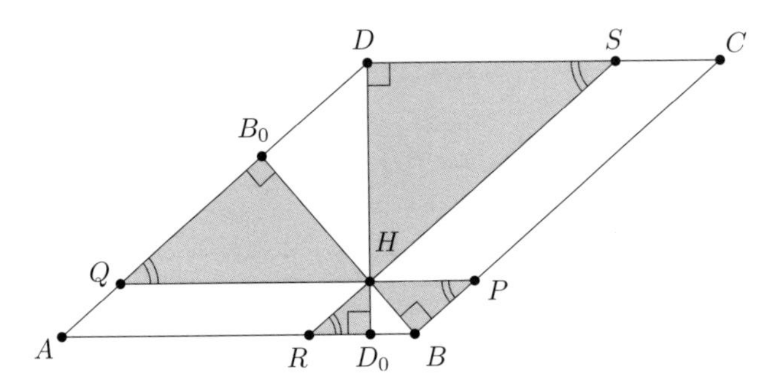

   Also, we observe that quadrilateral DB_0D_0B is cyclic (as $\angle BB_0D = 90^\circ = \angle DD_0B$) so by Power of a Point we have
   $$HB \cdot HB_0 = HD \cdot HD_0. \qquad (\star)$$

   Now we find similar triangles in order to link the two alike relations. Indeed, since the parallel lines give
   $$\angle BRH = \angle HPB = \angle DSH = \angle HQB_0,$$
   the four right triangles DHS, BHP, D_0HR, and B_0HQ are pairwise similar (AA). From here we learn
   $$\frac{HD_0}{HB_0} = \frac{HR}{HQ}, \quad \text{and} \quad \frac{HD}{HB} = \frac{HS}{HP}$$
   and after multiplying the two and comparing with $(\star)$, we obtain the result.

3. [Baltic Way 2010] Let ABC be an acute-angled triangle. Let D and E be points on the sides AB and AC such that B, C, D, and E lie on the same circle. Further, suppose the circle through D, E, and A intersects

the side BC in two points X and Y. Show that the midpoint of XY is the foot of the altitude from A to BC.

Proof. Let $A_0 \in BC$ be the foot of altitude from A. As BC and DE are antiparallel in $\angle BAC$ and AA_0 is altitude in triangle ABC, it must pass through the circumcenter of triangle ADE (H and O are friends – see Proposition 1.47).

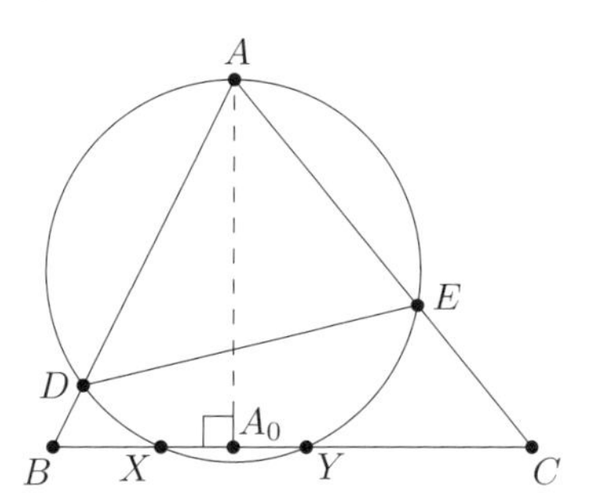

Thus, the circumcircle of triangle ADE is symmetric in line AA_0 and since X and Y correspond in this symmetry, A_0 is the midpoint of XY.

4. [Tournament of Towns 2007] Point B lies on a line which is tangent to circle ω at point A. The line segment AB is rotated about the center of the circle by some angle to form segment $A'B'$. Prove that the line AA' bisects the segment BB'.

First Proof. Without loss of generality assume that AA' is horizontal. Since the segments AB and $A'B'$ form the same angle with the line AA' (Equal Tangents) and are of equal length, point B is as much "above" AA' as B' is "below" it. Then the midpoint of BB' must lie on AA'.

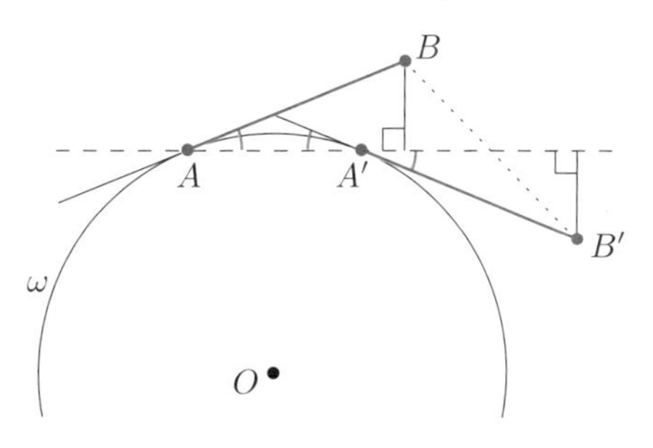

Second Proof. Let X be the intersection of AA' with BB' and Y the interesction of AB with $A'B'$. By Menelaus' Theorem in triangle YBB', we have that

$$\frac{BX}{XB'} \cdot \frac{B'A'}{A'Y} \cdot \frac{YA}{AB} = 1.$$

But $A'Y = AY$ (Equal Tangents) and $A'B' = AB$, so we immediately get that $BX = XB'$, which is precisely what we wanted.

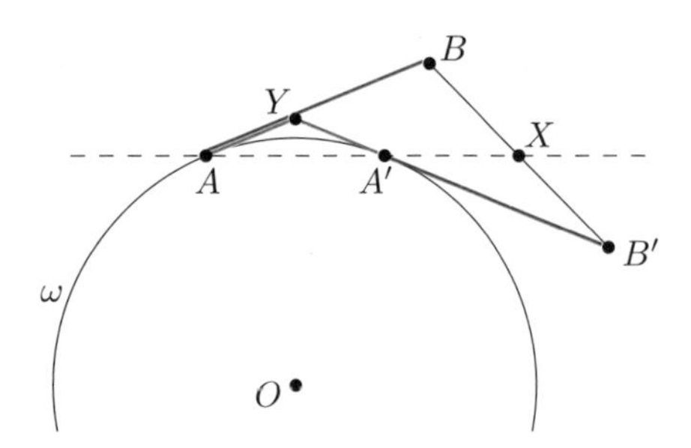

5. [USAMO 1998] Let ω_1 and ω_2 be concentric circles, with ω_2 in the interior of ω_1. From a point A on ω_1 draw the tangent AB to ω_2 ($B \in \omega_2$). Let C be the second point of intersection of AB and ω_1, and let D be the midpoint of AB. A line passing through A intersects ω_2 at E and F in such a way that the perpendicular bisectors of DE and CF intersect at a point M on AB. Find the ratio AM/MC.

Solution. We place AB horizontally and remove the circle ω_1 entirely keeping in mind that B is the midpoint of AC, which follows from symmetry. With what is left in the picture now, Power of a Point is a must.

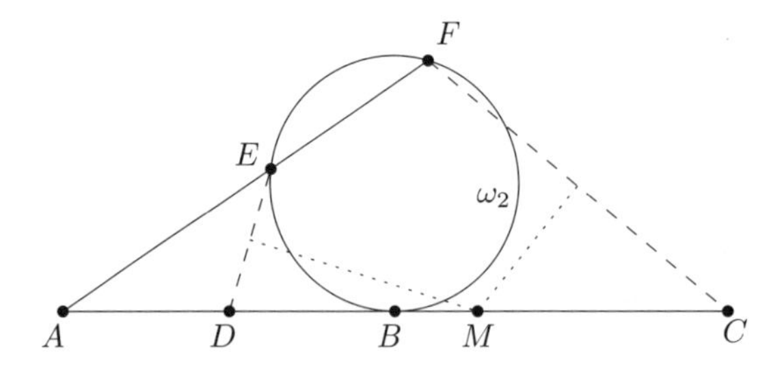

It yields

$$AE \cdot AF = AB^2 = \frac{AB}{2} \cdot 2AB = AD \cdot AC$$

implying that $EFCD$ is cyclic and thus M as the intersection of two perpendicular bisectors in $EFCD$ is inevitably its circumcenter. In particular, it is the midpoint of CD. Now we calculate the desired ratio easily as

A D B M C

$$AM = AD + \frac{1}{2}CD = \frac{1}{4}AC + \frac{3}{8}AC = \frac{5}{8}AC,$$

then $MC = \frac{3}{8}AC$ and the answer is $\frac{5}{3}$.

6. [Titu Andreescu] Let M be a point inside triangle ABC such that

$$AM \cdot BC + BM \cdot AC + CM \cdot AB = 4[ABC].$$

Show that M is the orthocenter of triangle ABC.

Proof. We aim to relate the products on the left-hand side to some areas.

Extend AM to meet BC at D and denote by B_0, C_0 the feet of perpendiculars dropped onto AM from B, C, respectively. Since perpendicular is the shortest distance from a point to a line, $BC = BD + DC \geq BB_0 + CC_0$ and

$$AM \cdot BC \geq AM \cdot BB_0 + AM \cdot CC_0 = 2 \cdot ([AMB] + [CMA]),$$

with equality if and only if $AM \perp BC$.

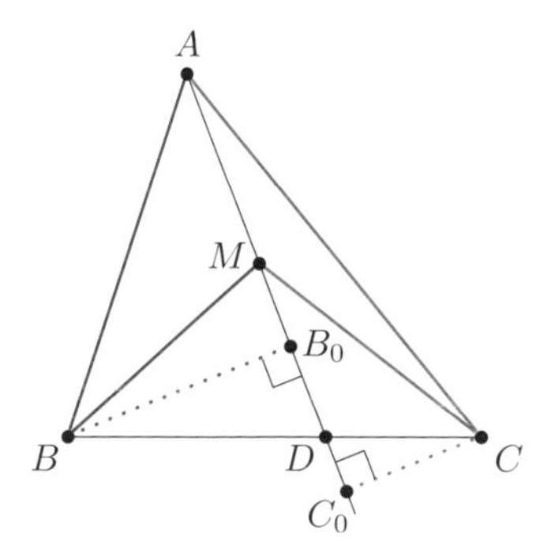

Likewise we obtain

$$BM \cdot AC \geq 2 \cdot ([BMC] + [AMB]),$$
$$CM \cdot AB \geq 2 \cdot ([CMA] + [BMC]).$$

Summing these three inequalities we get

$$AM \cdot BC + BM \cdot AC + CM \cdot AB \geq 4([AMB] + [BMC] + [CMA]) = 4[ABC].$$

Hence the equality in all three partial inequalities has to occur, implying that M is the orthocenter of triangle ABC.

7. Let ABC be a triangle. Prove that lines joining midpoints of the sides with midpoints of the corresponding altitudes pass through a single point.

Proof. Denote the midpoints of the sides BC, CA, AB by M, N, P, respectively, the feet of corresponding altitudes by D, E, F, and their midpoints by X, Y, Z, respectively.

Since X is the midpoint of AD, it lies on the midline NP of triangle ABC. Hence it is convenient to view the situation with respect to triangle MNP.

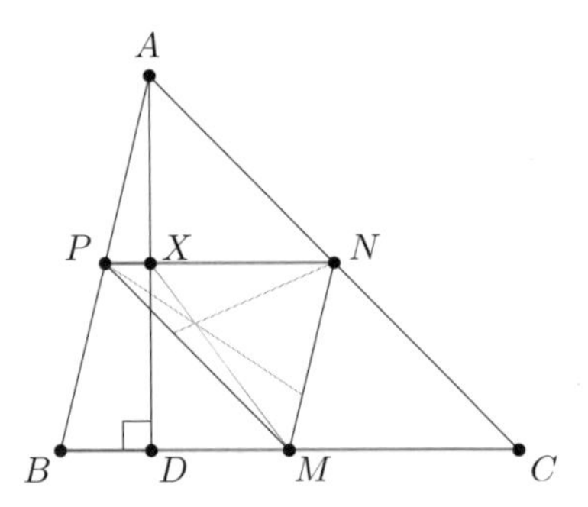

In order to establish the concurrence, by Ceva's Theorem it suffices to prove

$$\frac{NX}{XP} \cdot \frac{PY}{YM} \cdot \frac{MZ}{ZN} = 1.$$

Next we get rid of all the midpoints. As NX and XP are midlines in triangles CAD and DAB, we have $NX = \frac{1}{2}CD$ and $XP = \frac{1}{2}DB$, and thus the first ratio on the left-hand side equals CD/DB. By similar arguments we rewrite the other two ratios and learn

$$\frac{NX}{XP} \cdot \frac{PY}{YM} \cdot \frac{MZ}{ZN} = \frac{CD}{DB} \cdot \frac{AE}{EC} \cdot \frac{BF}{FA}$$

But the latter *is* equal to one by another Ceva's Theorem because the segments AD, BE, CF actually are concurrent cevians in triangle ABC (they concur at its orthocenter) so the proof is complete.

8. [Baltic Way 2011] Let $ABCD$ be a convex quadrilateral such that $\angle ADB = \angle BDC$. Suppose that a point E on the side AD satisfies the equality

$$AE \cdot ED + BE^2 = CD \cdot AE.$$

Show that $\angle EBA = \angle DCB$.

Proof. We draw DB vertically to be more aware of the symmetry. Observe that most distances in the given metric condition take place either on line DC or on the symmetric line DA. It should strike us that we could restate the problem by putting all distances just on one of the symmetric lines.

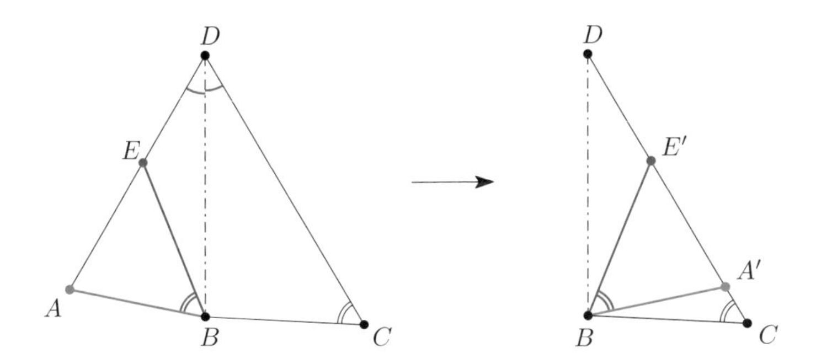

Indeed, let A' and E' be the reflections of A and E over BD. It's not a surprise that the condition simplifies:

$$BE'^2 = BE^2 = CD \cdot AE - AE \cdot ED = A'E'(CD - E'D) = A'E' \cdot CE'.$$

By Power of a Point, $E'B$ is tangent to the circumcircle of triangle $BA'C$ and so $\angle A'BE' = \angle DCB$ (see Proposition 1.34). The conclusion follows by symmetry.

9. [USAMO 2010] Let ABC be a triangle with $\angle A = 90°$. Denote its incenter by I and let $D = BI \cap AC$ and $E = CI \cap AB$. Determine whether or not it is possible for segments AB, AC, BI, ID, CI, IE to all have integer lengths.

   Proof. Having in mind basic angles in a triangle, we aim to make use of $\angle BIC = 135°$ (see Proposition 1.11). The trick is to express $\cos \angle BIC$ from the Law of Cosines in triangle BIC.

   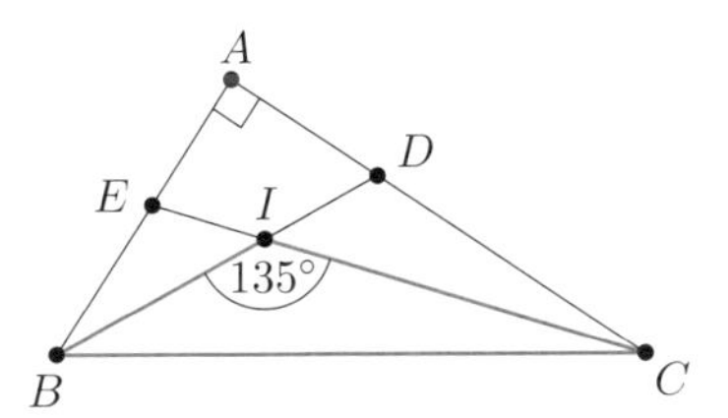

   With the help of the Pythagorean Theorem in triangle ABC, we obtain

$$-\frac{\sqrt{2}}{2} = \cos 135° = \frac{BI^2 + CI^2 - BC^2}{2BI \cdot CI} = \frac{BI^2 + CI^2 - AB^2 - AC^2}{2BI \cdot CI}.$$

   We see that the left-hand side is irrational, thus it is impossible for all the distances AB, AC, BI, and CI to have integer lengths.

10. Let A and B be two fixed points inside of the fixed circle ω symmetric with respect to its center O. If points M and N vary on ω in the same

half-plane with respect to AB, so that $AM \parallel BN$, prove that $AM \cdot BN$ is constant.

Proof. Extend MA to meet ω for the second time at N'. As A and B are symmetric about O and $AM \parallel BN$, points N and N' are also symmetric about O and $AN' = BN$. Thus the product $AM \cdot BN$ equals $AM \cdot AN'$ which is simply (negative) power of A with respect to ω, a constant.

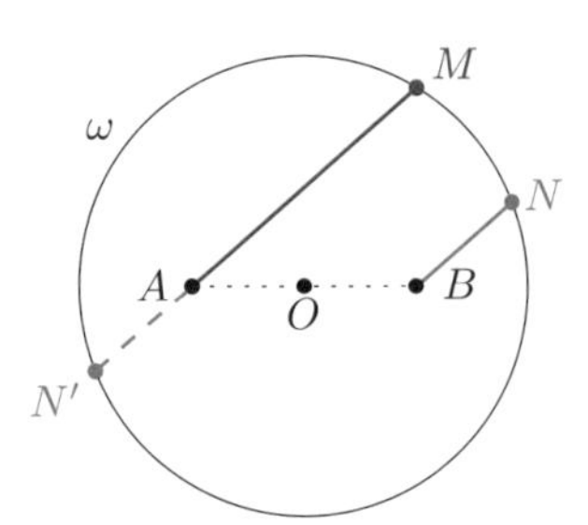

11. In a trapezoid $ABCD$, the segment connecting the midpoints M, N of the bases AB, CD, respectively, has length 4, and the diagonals have lengths $AC = 6$ and $BD = 8$ Find the area of the trapezoid.

    First Solution. Let L be the midpoint of BC. Then $ML = \frac{1}{2}AC = 3$ and $LN = \frac{1}{2}BD = 4$ so Heron's formula yields $[MLN] = \frac{3}{4}\sqrt{55}$. Next we relate $[MLN]$ to $[ABCD]$.

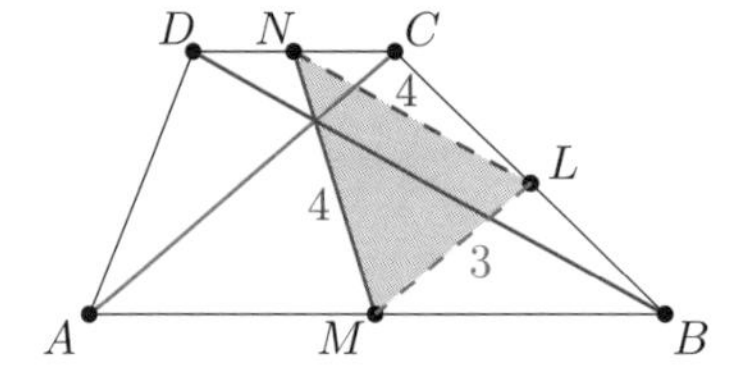

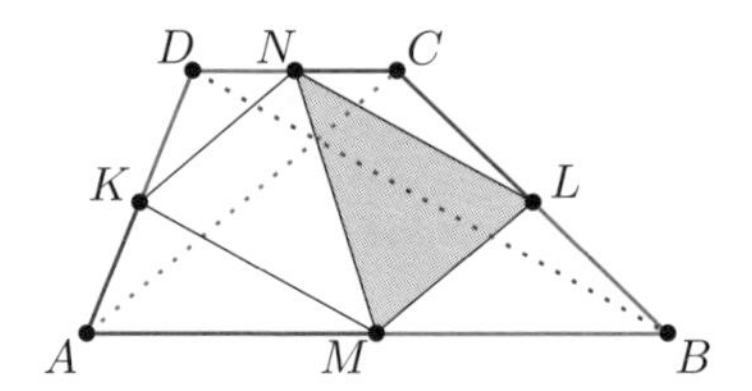

Let K be the midpoint of AD. As $KMLN$ is a parallelogram (see Introductory Problem 9), $[KMLN] = 2 \cdot [MLN]$. Regarding the areas of the remaining four small triangles, since ML and KN are midlines, we have

$$[MBL] + [KDN] = \frac{1}{4}[ABC] + \frac{1}{4}[ADC] = \frac{1}{4}[ABCD]$$

and likewise $[KAM] + [NCL] = \frac{1}{4}[ABCD]$. Thus the parallelogram $KMLN$ occupies $1 - \frac{1}{4} - \frac{1}{4} = \frac{1}{2}$ of the area of $ABCD$ and finally

$$[ABCD] = 2 \cdot [KMLN] = 4 \cdot [MLN] = 3\sqrt{55}.$$

Second Solution. As in the first solution, let K, L be the midpoints of the sides AD, BC, respectively.

We cut off the triangle LCN and flip it about L to obtain triangle LBZ (this is possible as $LB = LC$). Since $ABCD$ is a trapezoid, Z lies on the line AB, and we have $MZ = \frac{1}{2}AB + \frac{1}{2}CD$ and $NZ = 2 \cdot NL = DB = 8$.

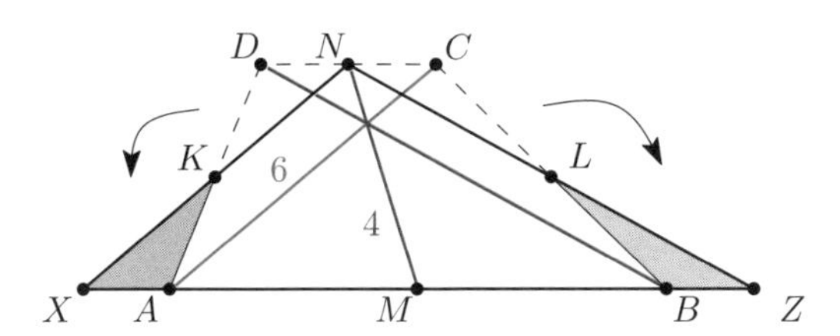

Doing the same for triangle KDN (i.e. flipping triangle KDN about K into triangle KAX) we realize that it suffices to determine the area of a triangle NXZ given the lengths of two of its sides NZ, NX, and the length of its median NM.

Let Y be the point such that $NXYZ$ is a parallelogram. Then $[NXZ] = \frac{1}{2}[NXYZ] = [NXY]$ and the side lengths of triangle NXY are known as $NX = 6$, $XY = NZ = 8$, and $NY = 2 \cdot NM = 8$. Hence we conclude using Heron's formula:

$$[ABCD] = [NXZ] = [NXY] = \sqrt{11 \cdot 5 \cdot 3 \cdot 3} = 3\sqrt{55}.$$

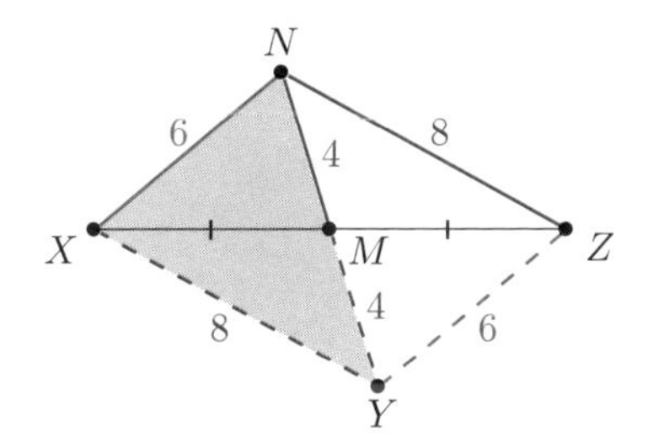

12. In triangle ABC, let AP, BQ, CR be concurrent cevians. Let the circumcircle of triangle PQR intersect the sides BC, CA, AB for the second time at X, Y, Z, respectively. Prove that AX, BY, CZ are concurrent.

Proof. We start with Ceva's Theorem for concurrent cevians AP, BQ, CR and we aim to apply another Ceva's Theorem for AX, BY, CZ. We expect to handle the circle by means of Power of a Point.

We have

$$AZ \cdot AR = AQ \cdot AY \quad \text{and thus} \quad \frac{AR}{AQ} = \frac{AY}{AZ}.$$

Likewise we obtain

$$\frac{BP}{BR} = \frac{BZ}{BX}, \qquad \frac{CQ}{CP} = \frac{CX}{CY}.$$

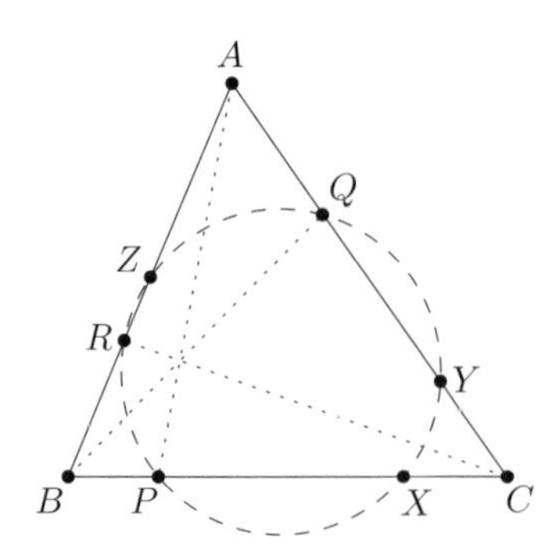

Now we choose to write the aforementioned Ceva's Theorem for AP, BQ, CR in the form

$$\frac{AR}{AQ} \cdot \frac{BP}{BR} \cdot \frac{CQ}{CP} = 1.$$

Substituting the derived equalities yields

$$\frac{AY}{AZ} \cdot \frac{BZ}{BX} \cdot \frac{CX}{CY} = 1,$$

which is equivalent to the desired metric condition from Ceva's Theorem for AX, BY, and CZ. We may conclude.

Remark. This is a particular case of a celebrated theorem by Carnot that if P, X lie on BC, Q, Y on CA, and R, Z on AB, then the points P, Q, R, X, Y, Z all lie on the same conic if and only if

$$\frac{XB}{XC} \cdot \frac{PB}{PC} \cdot \frac{YC}{YA} \cdot \frac{QC}{QA} \cdot \frac{ZA}{ZB} \cdot \frac{RA}{RB} = 1.$$

13. [All-Russian Olympiad 2002] A quadrilateral $ABCD$ is inscribed in a circle ω. The tangent to ω at B intersects the ray DC at K, and the tangent to ω at C intersects the ray AB at M. Prove that if $BM = BA$ and $CK = CD$, then $ABCD$ is a trapezoid.

First Proof. The midpoints suggest we should try to use concyclicity of $ABCD$ in terms of ratios.

By Power of a Point we learn $KB^2 = KC \cdot KD = 2KC^2$ and similarly $MC^2 = MB \cdot MA = 2MB^2$. Thus the ratios KB/KC and MC/MB are equal. Moreover, both angles $\angle BCM = \angle KBC$ correspond to the same arc BC and thus are also equal.

Now it is time to observe that triangles KCB and MBC have a lot in common. We exploit this by the Law of Sines:

$$\frac{KB}{KC} = \frac{\sin\angle KCB}{\sin\angle KBC}, \qquad \frac{MC}{MB} = \frac{\sin\angle MBC}{\sin\angle BCM},$$

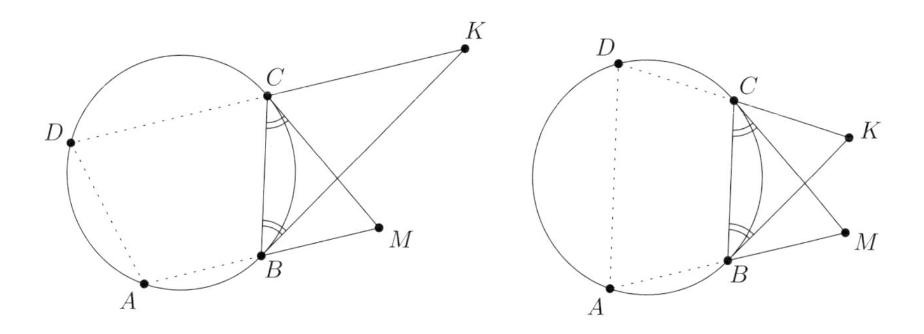

from which we learn $\sin \angle KCB = \sin \angle MBC$. We distinguish two cases. If $\angle KCB = \angle MBC$, then $BC \parallel DA$ and if $\angle KCB = 180^\circ - \angle MBC$, then $AB \parallel CD$. Either way, $ABCD$ is a trapezoid and we may conclude.

Second Proof. We approach the problem as a ruler-compass construction. Starting only with ω and points B and C, we are looking for suitable choices of A and D. Since B is to be the midpoint of AM, the point A must lie on line ℓ' which is symmetric in point B with the tangent ℓ to ω at C.

Let the two intersections of ℓ' and ω be A' and A'', the only possible positions of point A. Since $\ell \parallel \ell'$, point C is the midpoint of arc $A'A''$ (containing B) and hence $A'C = A''C$.

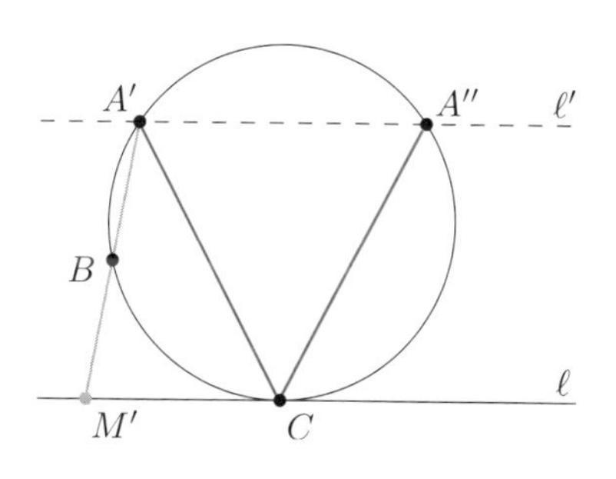

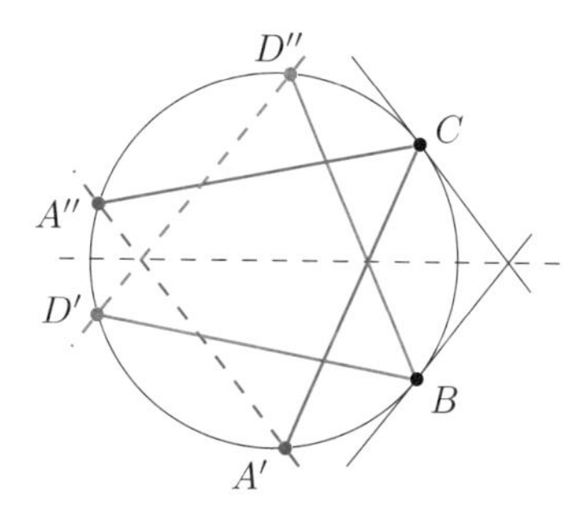

We have shown that the diagonal AC has equal length for both positions of A. The same holds for the diagonal BD. Moreover, these lengths are equal as we can see from line symmetry in the axis of BC. Thus, $ABCD$ has equal diagonals so it is indeed a trapezoid (if in doubt, see Example 1.7).

14. Let $ABCD$ be a parallelogram and M, N points on its sides AB, AD such that $\angle MCB = \angle DCN$. Let P, Q, R, and S be the midpoints of the segments AB, AD, NB, and MD, respectively. Show that P, Q, R, and S are concyclic.

Proof. Our first observation is that PR is the midline in triangle ABN

and QS is the midline in triangle ADM. These midlines are in fact the midlines of parallelogram $ABCD$ and intersect at its center O.

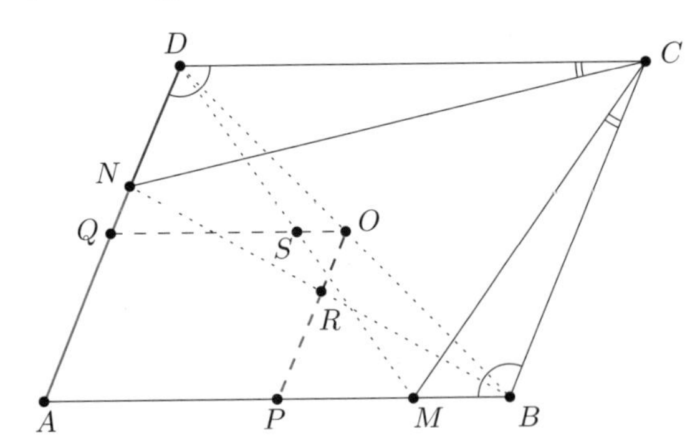

We will show concyclicity by means of Power of a Point. We want to prove

$$OR \cdot OP = OS \cdot OQ.$$

As OR and OP are midlines in triangles DBN and DBA, we see that $OR = \frac{1}{2}DN$ and $OP = \frac{1}{2}DA$. Similarly, we derive $OS = \frac{1}{2}BM$ and $OQ = \frac{1}{2}BA$. Now it suffices to prove

$$DN \cdot DA = BM \cdot BA \quad \text{or} \quad \frac{DN}{DC} = \frac{BM}{BC},$$

but the last relation follows from the similarity of triangles MBC and NDC (AA). We are done.

15. [Tournament of Towns 2008] Diagonals of non-isosceles trapezoid $ABCD$ intersect at P. Let A_1 be the second intersection of the circumcircle of triangle BCD and AP. Points B_1, C_1, D_1 are defined in a similar way. Prove that $A_1B_1C_1D_1$ is also a trapezoid.

    Proof. It suffices to consider the case when $AB \parallel CD$. Denote the lengths of the segments PA, PB, PC, PD by a, b, c, d, respectively. We intend to locate the positions of the points A_1, B_1, C_1, and D_1 on the diagonals AC and BD by Power of a Point and then to use similarity to work with the trapezoids.

    Taking power of P with respect to the four circles yields

$$PA_1 = \frac{bd}{c}, \quad PB_1 = \frac{ac}{d}, \quad PC_1 = \frac{bd}{a}, \quad \text{and} \quad PD_1 = \frac{ac}{b}.$$

From $AB \parallel CD$ we infer $\triangle ABP \sim \triangle CDP$ and $b/a = d/c$. Hence equations

$$\frac{PA_1}{PB_1} = \frac{bd}{c} \cdot \frac{d}{ac} = \frac{bd}{ac} \cdot \frac{d}{c} \quad \text{and} \quad \frac{PC_1}{PD_1} = \frac{bd}{a} \cdot \frac{b}{ac} = \frac{bd}{ac} \cdot \frac{b}{a}$$

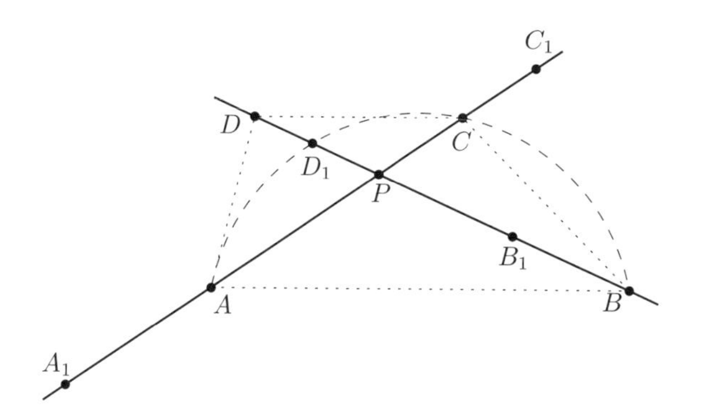

imply that $PA_1/PB_1 = PC_1/PD_1$. As a consequence, the triangles A_1B_1P and C_1B_1D are similar (SAS) and $A_1B_1C_1D_1$ is indeed a trapezoid.

16. [Czech and Slovak 2006] Let ω be a circle with center O and radius r and A a point different from O. Find the locus of circumcenters of the triangles ABC for which BC is a diameter of ω.

    Solution. Here the key is the following observation. The point O has constant power with respect to all circumcircles of ABC, namely $-OB \cdot OC = -r^2$. Then the second intersection X of line AO with the circumcircle of ABC is fixed, since from

$$OA \cdot OX = OB \cdot OC$$

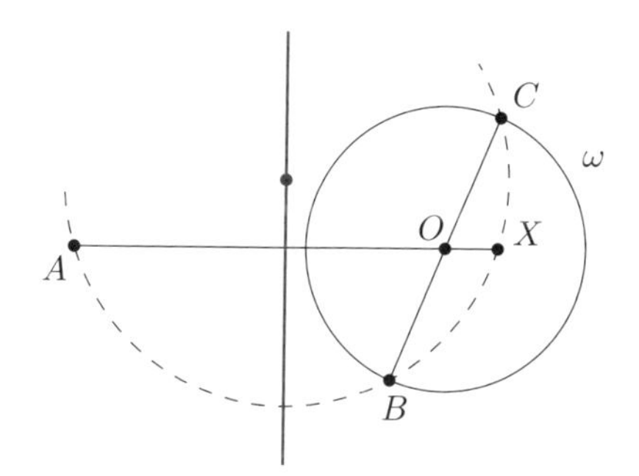

we infer $OX = r^2/OA$. Therefore all the circles pass through two fixed points, hence their centers lie on a fixed line (namely the perpendicular bisector of AX). In order to show that any point of this line belongs to the desired locus, it suffices to note that any circle through A and X intersects ω at two diametrically opposite points, which can be seen from reversing the performed calculations.

17. Let $ABCD$ be quadrilateral such that

$$\angle ADB + \angle ACB = 90^\circ \quad \text{and} \quad \angle DBC + 2\angle DBA = 180^\circ.$$

Show that

$$(DB + BC)^2 = AD^2 + AC^2.$$

Proof. The strange angular conditions and the form of the conclusion resembling the Pythagorean Theorem call for some transformation. First, we erase the segment CD. Then we look at the figure as if it was folded along the line AB. Unfolding does the trick!

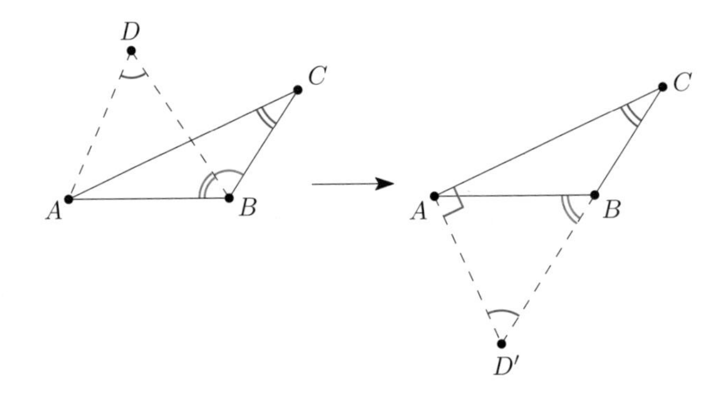

Indeed, if D' is the mirror image of D over AB, then

$$\angle CBD' = \angle CBD + 2 \cdot \angle DBA = 180^\circ,$$

and the points C, B, and D' are collinear. Also, in triangle $AD'C$ we have $\angle D'AC = 180^\circ - \angle ACB - \angle BD'A = 90^\circ$, hence it is right. The conclusion now follows from the Pythagorean Theorem (and symmetry).

18. [Poland 2008] We are given a triangle ABC such that $AB = AC$. There is a point D lying on the segment BC, such that $BD < DC$. Point E is symmetrical to B with respect to AD. Prove that

$$\frac{AB}{AD} = \frac{CE}{CD - BD}.$$

First Proof. Let D' be the point on segment BC with $D'C = BD$. Then $DD' = CD - D'C = CD - BD$. By symmetry we have $\angle BAD = \angle D'AC$ and if we also use the symmetry about AD, we learn $\angle DAD' = \angle BAC - 2\angle BAD = \angle EAC$.

Thus triangles $D'AD$ and EAC being isosceles with the same vertex angle are similar. Hence

$$\frac{AB}{AD} = \frac{AC}{AD'} = \frac{CE}{DD'} = \frac{CE}{CD - BD}$$

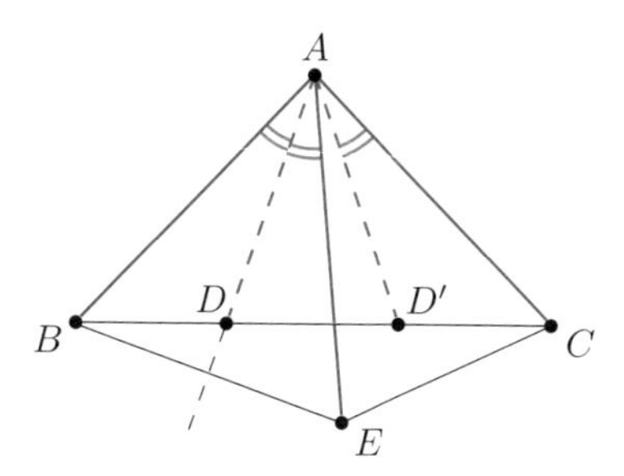

and we are done.

Second Proof. We construct the midpoint M of BC and denote by F the intersection of AD and BE.

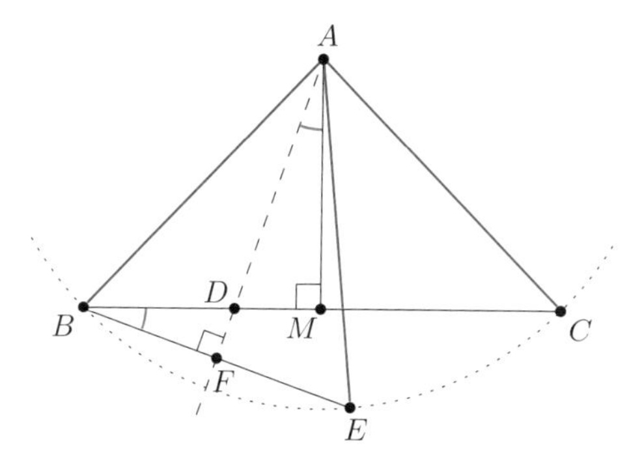

From similar right triangles DFB and DMA (AA), we have $\angle FBM = \angle FAM$. Next, simple calculation gives

$$CD - BD = (CM + MD) - (BM - MD) = 2MD.$$

Last thing to observe is that $AC = AB = AE$ (the last equality is due to symmetry), hence A is the circumcenter of triangle BCE. Then from the Extended Law of Sines in this triangle we learn that $CE = 2AB \cdot \sin \angle EBC$. Now putting all this together yields the desired

$$\frac{AB}{AD} = \frac{AB}{MD} \cdot \sin \angle FAM = \frac{2 \cdot AB \cdot \sin \angle FBM}{2MD} = \frac{CE}{CD - BD}.$$

Third Proof. This time we introduce point D' on the segment BC such that $CD' = CD - BD$. Then D is the midpoint of BD'. We aim to prove $\triangle ABD \sim \triangle CED'$.

As before we observe that A is the circumcenter of triangle BEC and deduce that $\angle BAD = \angle BCE$ since both are equal to one half of the central angle BAE. Moreover, as the line AD intersects both segments BD' and BE at their midpoints, it is in fact the midline of triangle

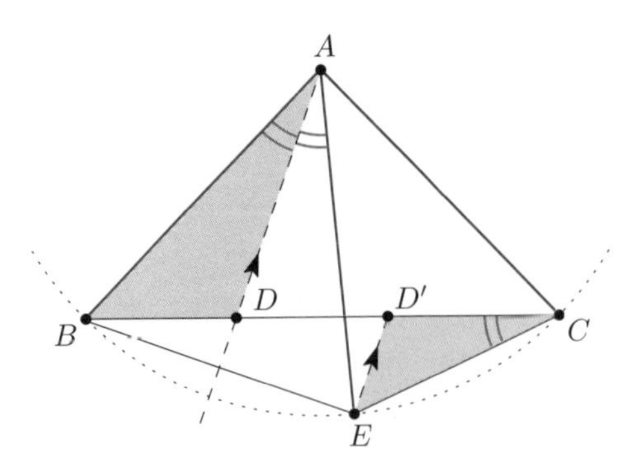

$BD'E$, thus $AD \parallel D'E$. Then $\angle ADB = \angle ED'C$, so now we indeed have $\triangle ABD \sim \triangle CED'$ (AA), which yields the desired

$$\frac{AB}{AD} = \frac{CE}{CD'} = \frac{CE}{CD - BD}.$$

19. Let P be a point on the side BC of triangle ABC. Perpendicular bisectors of the sides AB and AC meet the segment AP at points D and E, respectively. The line parallel to AB passing through D intersects the tangent to the circumcircle ω of triangle ABC through B at point M. Similarly, the line parallel to AC passing through E intersects the tangent to ω through C at point N. Prove that MN is tangent to ω.

Proof. Everything seems to be arranged for angle-chasing. As usual, we do not draw the perpendicular bisectors and work with isosceles triangles BDA and CEA instead. First, we focus on the left part of the picture. From the isosceles triangle BDA we have $\angle BAD = \angle DBA$, then as $DM \parallel AB$, also $\angle BDM = \angle DBA$. Next, $\angle BDP$ is external angle in triangle BDA, thus $\angle BDP = 2\angle BAD$, and so $\angle MDP = \angle BDP - \angle BDM = \angle BAD$.

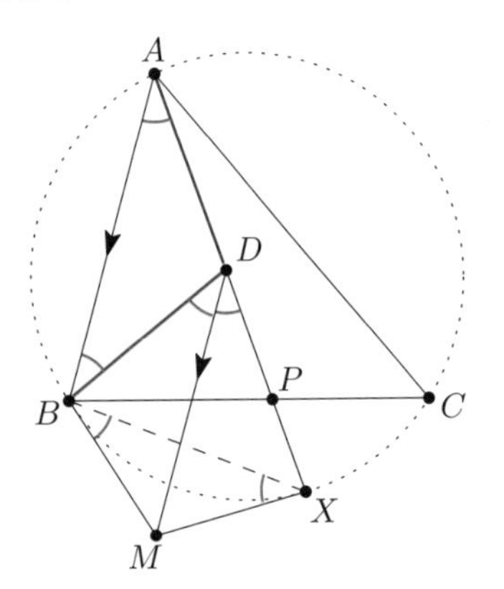

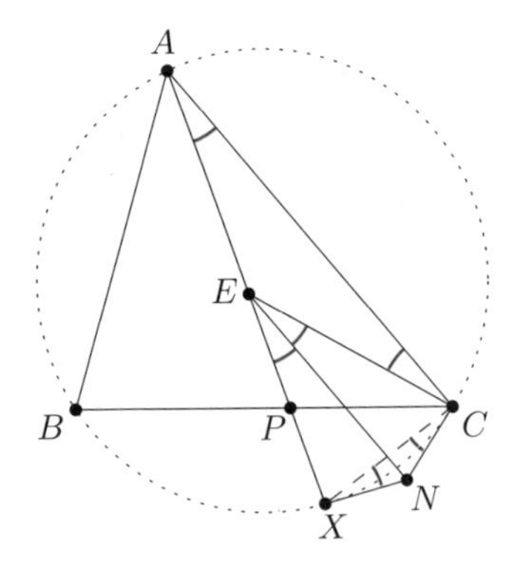

Now comes the vital step. Our common angle chasing tools can only prove that a line is tangent at a certain point. But here, we have no information about the point of contact. We must guess where it is. With good intuition or some skill in mental angle-chasing, we choose the second intersection X of AP and ω. Then since BM is tangent to ω we have $\angle MBX = \angle BAX$ (see Proposition 1.34), and so $\angle MBX = \angle MDX$ therefore $BMXD$ is cyclic. Finally, we deduce that $\angle BXM = \angle BDM = \angle BAP$ which implies that MX is tangent to ω.

Analogously, we prove that NX is tangent to ω, therefore MN is a tangent to ω at X.

20. In an acute triangle ABC a semicircle ω centered on the side BC is tangent to the sides AB and AC at points F and E, respectively. If X is the intersection of BE and CF, show that $AX \perp BC$.

Proof. If we denote by D the foot of the A-altitude, we may change the perspective and switch to proving that cevians AD, BE, and CF concur. By Ceva's Theorem we need to verify

$$\frac{BD}{DC} \cdot \frac{CE}{EA} \cdot \frac{AF}{FB} = 1,$$

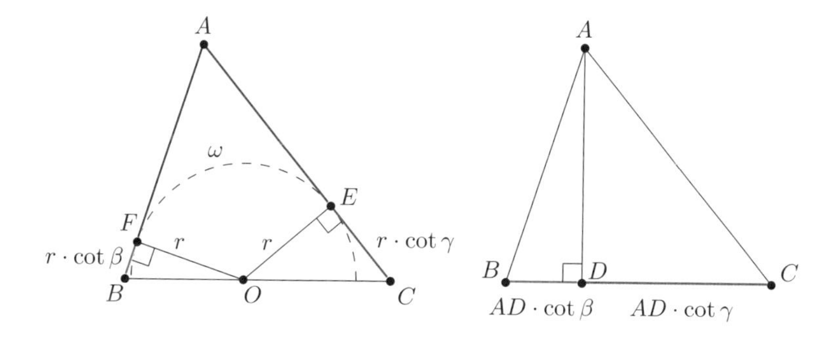

which since $AE = AF$ (Equal Tangents) reduces to

$$\frac{BD}{DC} = \frac{FB}{CE},$$

Now we denote the center of ω by O and its radius by r. Right triangles OFB and OEC then give

$$\frac{FB}{CE} = \frac{r \cdot \cot\beta}{r \cdot \cot\gamma} = \frac{\cot\beta}{\cot\gamma}$$

and similarly right triangles BDA and DCA give

$$\frac{BD}{DC} = \frac{AD \cdot \cot\beta}{AD \cdot \cot\gamma} = \frac{\cot\beta}{\cot\gamma}$$

which ends the proof.

21. Let $ABCD$ be a convex quadrilateral and X a point in its interior. Denote by ω_A the circle tangent to the sides AB and AD and passing through X. Define circles ω_B, ω_C, and ω_D similarly. Given that all these circles have equal radii, show that $ABCD$ is cyclic.

Proof. We denote the centers of ω_A, ω_B, ω_C, and ω_D, by O_A, O_B, O_C, and O_D, respectively. Now since the circles have equal radii and all pass through X, we have $O_AX = O_BX = O_CX = O_DX$, thus points O_A, O_B, O_C, and O_D lie on a circle with center X.

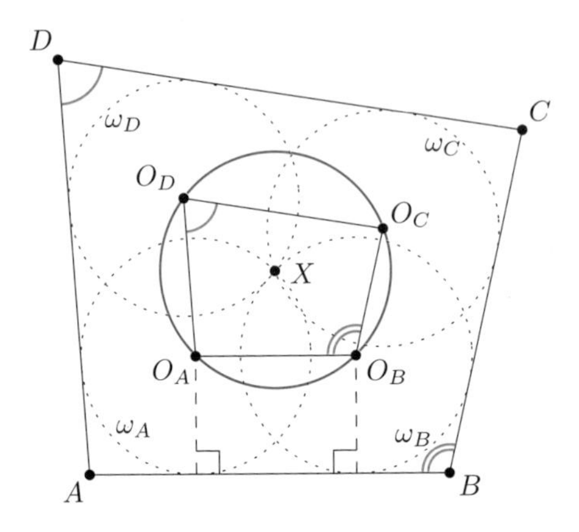

It remains to observe that $O_AO_B \parallel AB$, since points O_A, O_B have equal distance from the line AB. Similarly, we have $O_BO_C \parallel BC$, $O_CO_D \parallel CD$, and $O_DO_A \parallel DA$. Now it can be easily seen that

$$\angle CBA = \angle O_CO_BO_A = 180^\circ - \angle O_AO_DO_C = 180^\circ - \angle ADC,$$

hence $ABCD$ is cyclic as desired.

22. [Mathematical Reflections, Ivan Borsenco] In triangle ABC, let AP, BQ, CR be concurrent cevians. Denote by X, Y, Z the midpoints of segments QR, RP, PQ, respectively. Prove that the lines AX, BY, CZ are concurrent.

Proof. We intend to use the trigonometric form of Ceva's Theorem. From the Ratio Lemma (see Proposition 1.18) for triangle ARQ we learn

$$\frac{AR \sin \angle RAX}{QA \sin \angle XAQ} = \frac{RX}{XQ} = 1,$$

hence

$$\frac{\sin \angle RAX}{\sin \angle XAQ} = \frac{QA}{AR}.$$

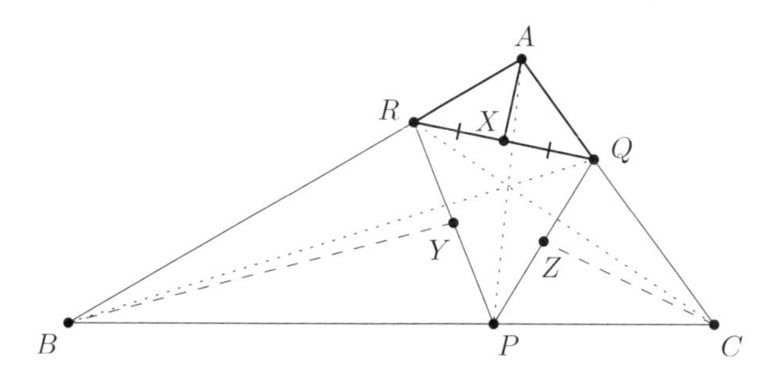

In the same vein, we find two analogous relations and we may write

$$\frac{\sin \angle RAX}{\sin \angle XAQ} \cdot \frac{\sin \angle PBY}{\sin \angle YBR} \cdot \frac{\sin \angle QCZ}{\sin \angle ZCP} = \frac{QA}{AR} \cdot \frac{RB}{BP} \cdot \frac{PC}{CQ},$$

where the right-hand side is equal to 1 by Ceva's Theorem applied to concurrent cevians AP, BQ, and CR. Thus also the left-hand side equals 1 and we are done by the trigonometric form of Ceva's Theorem.

Remark. The conclusion of the theorem still holds if we take any triplet of points X, Y, Z on the sides QR, RP, PQ of triangle PQR for which the lines PX, QY, RZ concur. The result is called the *Cevian Nest Theorem* and is proved with the same technique as this problem.

23. [USAJMO 2012] Given a triangle ABC, let P and Q be points on segments AB and AC, respectively, such that $AP = AQ$. Let S and R be distinct points on segment BC such that S lies between B and R, $\angle BPS = \angle PRS$, and $\angle CQR = \angle QSR$. Prove that P, Q, R, S are concyclic.

Proof. We interpret the angle relations as tangency. From $\angle BPS = \angle PRS$ we infer that BP is tangent to the circumcircle ω_1 of triangle SRP and $\angle CQR = \angle QSR$ implies CQ is tangent to the circumcircle ω_2 of triangle RQC (see Proposition 1.34). We are asked to prove that circles ω_1 and ω_2 coincide. Let us assume they are distinct.

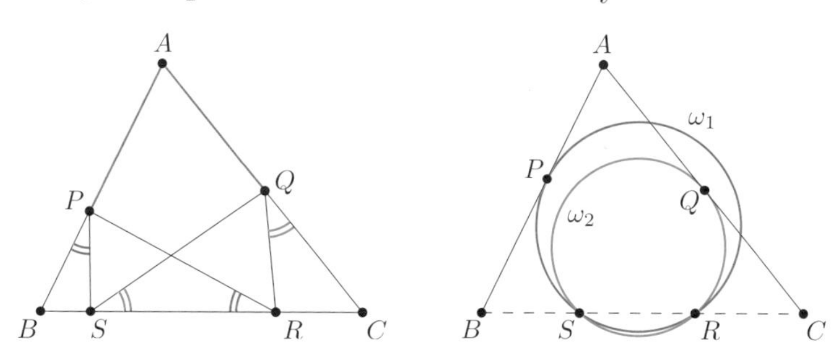

Then their radical axis is the line SR and at the same time

$$p(A, \omega_1) = AP^2 = AQ^2 = p(A, \omega_2).$$

Since $A \notin BC$, we reached a contradiction.

24. Segment AT is tangent to circle ω at T. A line parallel to AT intersects ω at B, C (with $AB < AC$). Lines AB, AC intersect ω for the second time at P, Q. Prove that line PQ bisects segment AT.

    Proof. Suppose that points A, B, P, and A, Q, C are collinear in this order (the other cases are analogous). Since $AT \parallel BC$ and $BQCP$ is cyclic, we have

$$\angle TAC = \angle BCA \equiv \angle BCQ = \angle BPQ.$$

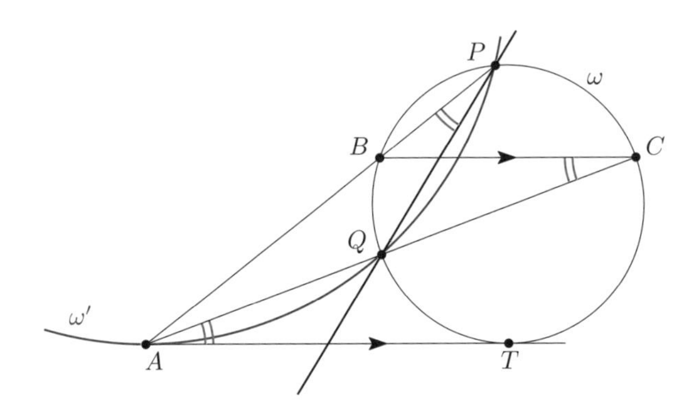

This implies that AT is tangent to the circumcircle ω' of triangle APQ (see Proposition 1.34). And now we are done because the radical axis PQ of ω and ω' bisects their common tangent AT (see Proposition 1.41).

25. [China 1990] Diagonals AC and BD of a cyclic quadrilateral $ABCD$ meet at P. Let the circumcenters of $ABCD$, ABP, BCP, CDP, and DAP be O, O_1, O_2, O_3, and O_4, respectively. Prove that OP, O_1O_3, and O_2O_4 are concurrent.

    Proof. We claim that PO_1OO_3 is a parallelogram. First, since AB and CD are antiparallel in $\angle APB$ and since "H and O are friends" (see Proposition 1.47), the line O_1P is also the altitude in triangle CPD, thus $O_1P \perp CD$.

    But also $OO_3 \perp CD$, since both circumcenters lie on the perpendicular bisector of CD. Hence $OO_3 \parallel O_1P$. Similarly, we prove that $OO_1 \parallel O_3P$ and so PO_1OO_3 is indeed a parallelogram.

    For analogous reasons OO_2PO_4 is also a parallelogram and since diagonals of a parallelogram bisect each other, the lines OP, O_1O_3, and O_2O_4 are concurrent at the midpoint of OP.

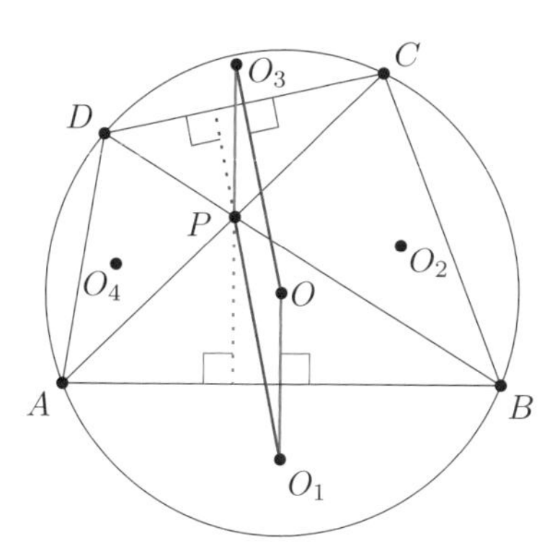

26. [AIME 2005] Triangle ABC has $BC = 20$. The incircle of the triangle evenly trisects the median AD at points E and F. Find the area of the triangle.

First Solution. As an A-isosceles triangle certainly does not have the property, we may assume $b > c$. Also let the incircle touch the sides BC, CA, AB at points X, Y, and Z, respectively.

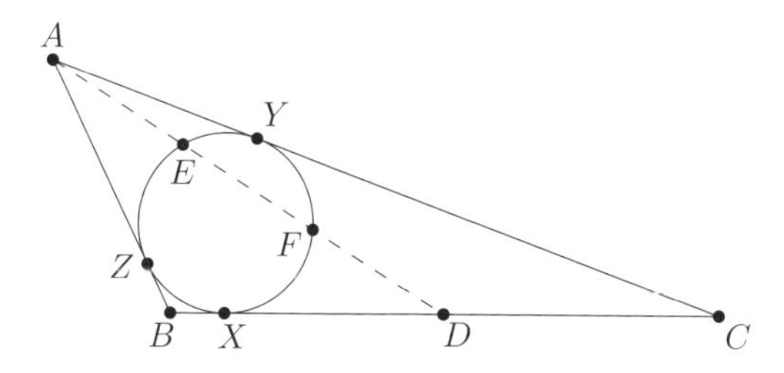

We will calculate the triangle side lengths with the help of Power of a Point, but first we recall basic distances in a triangle (see Proposition 1.15(a)) and find that

$$AZ = \frac{b+c-a}{2}, \quad DX = \frac{a}{2} - BX = \frac{a}{2} - \frac{a+c-b}{2} = \frac{b-c}{2}.$$

Taking powers of A and D with respect to the incircle we learn

$$AE \cdot AF = AZ^2 \quad \text{or} \quad \left(\frac{1}{3}AD\right)\left(\frac{2}{3}AD\right) = \left(\frac{b+c-a}{2}\right)^2,$$

and likewise

$$DF \cdot DE = DX^2 \quad \text{or} \quad \left(\frac{1}{3}AD\right)\left(\frac{2}{3}AD\right) = \left(\frac{b-c}{2}\right)^2.$$

As the left-hand sides are equal, we may compare the right-hand sides and find that $2c = a$ i.e. $c = 10$. Next, we use the median formula (see

Corollary 1.24) for AD and plugging it in we set a quadratic equation for b:

$$\frac{2}{9}\cdot\left(\frac{b^2+10^2}{2}-\frac{20^2}{4}\right)=\left(\frac{b-10}{2}\right)^2.$$

Simplifying we get $b^2-36b+260=0$ with solution $b=26$ (and also $b=10$ but this does not satisfy $b>c$).

Finally, we calculate the area K from Heron's formula

$$K=\sqrt{8\cdot 2\cdot 18\cdot 28}=24\sqrt{14}.$$

Second Solution. First, let us draw the median AD and the incircle ω only. Since the median is trisected evenly, the whole figure is symmetric about the perpendicular bisector of AD. Hence the tangents to ω issued from A and D, which touch ω in the same half-plane determined by AD, form an isosceles triangle.

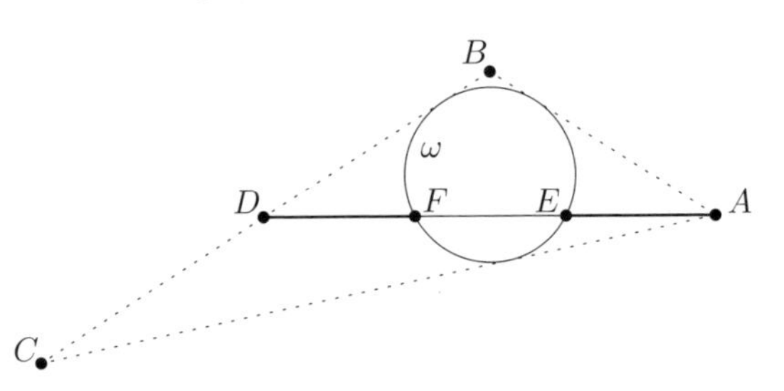

It remains to observe that the intersection of these tangents coincides with one of the vertices B, C of the triangle ABC. If we without loss of generality assume that it is B, we obtain $AB=BD=\frac{1}{2}BC$. We continue as in the first solution.

27. [IMO 1996 shortlist, Titu Andreescu] Let P be a point inside an equilateral triangle ABC. Let the lines AP, BP, CP meet the sides BC, CA, AB at the points A_1, B_1, C_1, respectively. Prove that

$$A_1B_1\cdot B_1C_1\cdot C_1A_1\geq A_1B\cdot B_1C\cdot C_1A.$$

Proof. Ceva's Theorem for concurrent lines AA_1, BB_1, CC_1 gives

$$A_1B\cdot B_1C\cdot C_1A=A_1C\cdot B_1A\cdot C_1B$$

which allows us to be proving a more symmetrical statement. Indeed, plugging it in the square of the sought-after inequality shows that it suffices to prove

$$(A_1B_1\cdot B_1C_1\cdot C_1A_1)^2\geq A_1B\cdot A_1C\cdot B_1A\cdot B_1C\cdot C_1A\cdot C_1B.$$

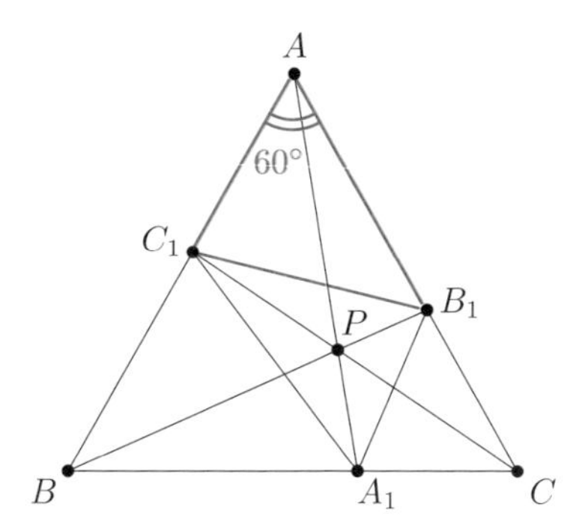

Applying the Law of Cosines to triangle AB_1C_1 we obtain

$$B_1C_1^2 = C_1A^2 + B_1A^2 - C_1A \cdot B_1A$$

and using the inequality $x^2 + y^2 - xy \geq xy$, which holds for all real numbers x, y, we get

$$B_1C_1^2 \geq C_1A \cdot B_1A.$$

After obtaining two analogous inequalities and multiplying all three, we get the result. Equality holds if and only if $CA_1 = CB_1$, $AB_1 = AC_1$, and $BC_1 = BA_1$, which in turn holds if and only if P is the center of the triangle ABC.

28. Let P and Q be isogonal conjugates[1] with respect to the triangle ABC. Show that the six feet of perpendiculars from P and Q to the sides of triangle ABC lie on one circle.

    Denote the feet from P to BC, CA, AB by P_a, P_b, and P_c, respectively, and define Q_a, Q_b, Q_c analogously.

    First Proof. Let $\angle BAP = \angle QAC = \varphi$ and $\angle BAQ = \angle PAC = \psi$. First we use Power of a Point to show that P_c, Q_c, P_b, Q_b are concyclic.

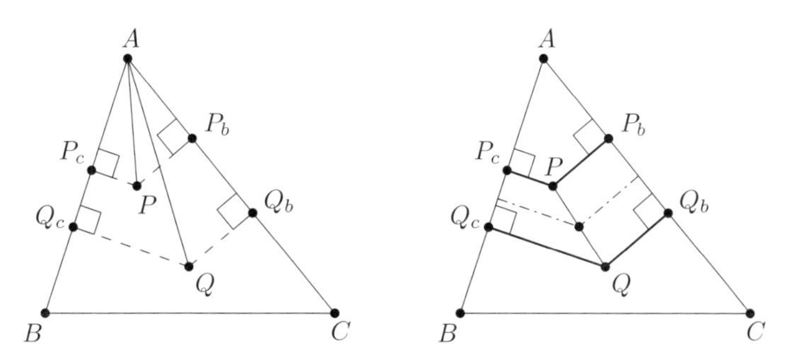

From right triangles APP_c and AQQ_c we have

$$AP_c \cdot AQ_c = (AP \cos \varphi) \cdot (AQ \cos \psi)$$

[1] For explanation see Theorem 1.46.

and similarly from APP_b and AQQ_b we obtain

$$AP_b \cdot AQ_b = (AP\cos\psi)\cdot(AQ\cos\varphi).$$

Since the products on the right-hand sides are equal, points P_c, Q_c, P_b, Q_b lie on a single circle.

Moreover, the center of this circle is the intersection of perpendicular bisectors of P_cQ_c and P_bQ_b, which are both midlines in trapezoids P_cQ_cQP and P_bQ_bQP and thus meet at the midpoint of PQ.

Analogously, we can show that P_a, Q_a, P_c, Q_c lie on a circle centered at the midpoint of PQ. The two circles share a center and two points on their perimeters and thus coincide. All six feet then lie on one circle.

Second Proof. As lines AP and AQ are isogonal in $\angle A$ and AP (being the diameter of the circumcircle of AP_cPP_b) passes through the circumcenter of AP_cP_b, it follows (H and O are friends – see Proposition 1.47) that $AQ \perp P_bP_c$. Similarly, we can show that $AP \perp Q_bQ_c$. Finally, since AP and AQ are antiparallel in $\angle A$, the perpendicular lines, namely P_bP_c and Q_bQ_c, are also antiparallel. Hence P_c, Q_c, P_b, Q_b lie on a single circle (call it ω_a).

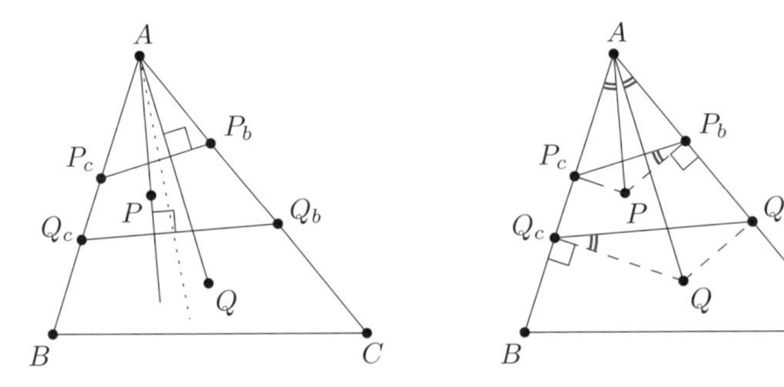

Now we could join the first proof but we proceed differently. As above we deduce that points P_a, Q_a, P_c, Q_c lie on some circle ω_b and that points P_a, Q_a, P_b, Q_b lie on some circle ω_c. Can these three circles be mutually different? No, since their pairwise radical axes would in that case be the lines AB, BC, CA which are neither parallel nor concurrent (a contradiction with Proposition 1.42). Hence at least two of these three circles coincide implying that all six points lie on a single circle.

Third Proof. Direct angle-chasing is possible but in order to avoid casework caused by undetermined order of points on the triangle sides, it is convenient to use directed angles. From cyclic quadrilaterals AP_cPP_b and AQ_cQQ_b we infer

$$\angle(P_cP_b, P_bQ_b) = \angle(P_cP_b, P_bP) + 90^\circ = \angle(P_CA, AP) + 90^\circ,$$
$$\angle(P_cQ_C, Q_cQ_b) = 90^\circ + \angle(QQ_c, Q_CQ) = 90^\circ + \angle(QA, AQ_b),$$

and thus $\angle(P_cP_b, P_bQ_b) = \angle(P_cQ_c, Q_cQ_b)$. The concyclicity of P_c, Q_c, P_b, Q_b follows and we may join any of the first two proofs.

Remark. If we take the isogonal conjugates to be H and O (orthocenter and circumcenter) the circle we obtain is the well-known *nine-point circle.* More about the nine-point circle will be discussed in the sequel to this book *107 Geometry Problems from the AwesomeMath Year-Round Program.*

29. The incircle of triangle ABC is tangent to its sides BC, CA, AB at points D, E, F, respectively. The excircles of triangle ABC are tangent to the corresponding sides of triangle ABC at points T, U, V. Show that triangles DEF and TUV have the same area.

Proof. First, we recall that $AF = BV = x$, $BD = CT = y$, and $CE = AU = z$ (see Proposition 1.15(a), (c)). Then the idea is to express both areas in terms of x, y, z and then just check an algebraic equality.

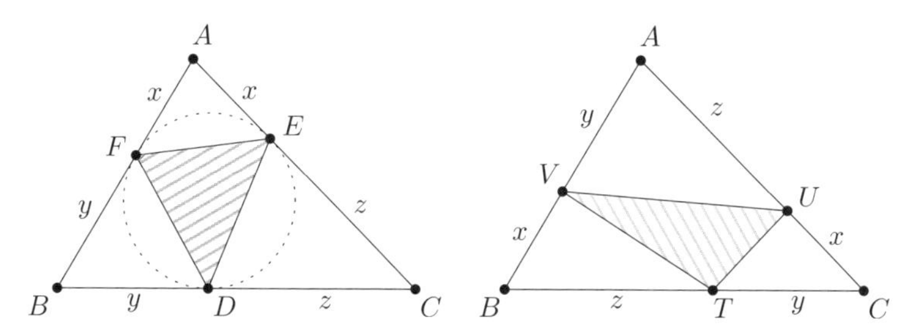

In fact, it suffices to compare the complements in triangle ABC

$$[ABC] - [DEF] = [AFE] + [BDF] + [CED],$$
$$[ABC] - [TUV] = [AVU] + [BTV] + [CUT].$$

We express the areas with area formula $2K = bc \sin\angle A$ and the Extended Law of Sines in the form $\sin\angle A = a/2R = (y+z)/2R$, where R is the circumradius of triangle ABC. We obtain

$$\begin{aligned}[AFE] + [BDF] + [CED] &= \frac{1}{2}\left(x^2 \sin\angle A + y^2 \sin\angle B + z^2 \sin\angle C\right) \\ &= \frac{1}{4R}\left(x^2(y+z) + y^2(z+x) + z^2(x+y)\right)\end{aligned}$$

and

$$\begin{aligned}[AVU] + [BTV] + [CUT] &= \frac{1}{2}\left(yz \sin\angle A + zx \sin\angle B + xy \sin\angle C\right) \\ &= \frac{1}{4R}\big(yz(y+z) + zx(z+x) + xy(x+y)\big).\end{aligned}$$

Since the complements are indeed equal, we may conclude.

30. [IMO 2008] Let H be the orthocenter of an acute-angled triangle ABC. The circle Γ_A centered at the midpoint of BC and passing through H intersects the sideline BC at points A_1 and A_2. Similarly, define the points B_1, B_2, C_1, and C_2. Prove that six points A_1, A_2, B_1, B_2, C_1, and C_2 are concyclic.

Proof. First, we will prove that the points B_1, B_2, C_1, and C_2 are concyclic. By the Radical Lemma (see Proposition 1.43) this is the case if and only if the radical axis of Γ_B, Γ_C passes through A. Denote the midpoints of the sides AB, AC by M, N, respectively.

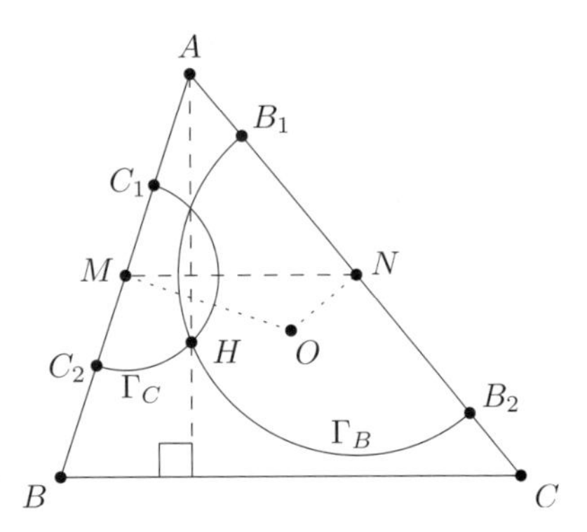

The radical axis of Γ_B and Γ_C is perpendicular to MN and hence also to BC. As it passes through H, it is the A-altitude in triangle ABC. Thus, it passes through A as well and the points B_1, B_2, C_1, and C_2 lie on one circle indeed. The center of this circle is the intersection of the perpendicular bisectors of B_1B_2 and C_1C_2, i.e. the circumcenter O of triangle ABC.

By symmetry, A_1, A_2, B_1, and B_2 lie on a circle centered at O as well. As both these circles pass through B_1 and B_2, they in fact coincide and the claim is proved.

31. [based on Sharygin Geometry Olympiad 2011] Distinct points A, B are given in the plane. Determine the locus of points C such that in triangle ABC the length of A-altitude is the same as the length of B-median.

Solution. Suppose we found point C such that if we denote the lengths of B-median BM and A-altitude AD of triangle ABC by m_b, h_a, respectively, then $m_b = h_a$. We intend to relate AD and BM.

To this end, let X be the point such that B is the midpoint of AX. Then BM is the midline in triangle AXC so $XC = 2 \cdot m_b$. Also, if we denote the foot of perpendicular dropped from X onto BC by D' then the triangles ABD and XBD' are congruent so $XD' = h_a$. Hence in the right triangle $CD'X$ we know $\sin\angle D'CX = \frac{1}{2}$ so $\angle D'CX = 30^\circ$ and thus $\angle BCX$ is either 30° or 150° (as it may be either congruent or supplementary to $\angle D'CX$, depending on the position of D' on BC).

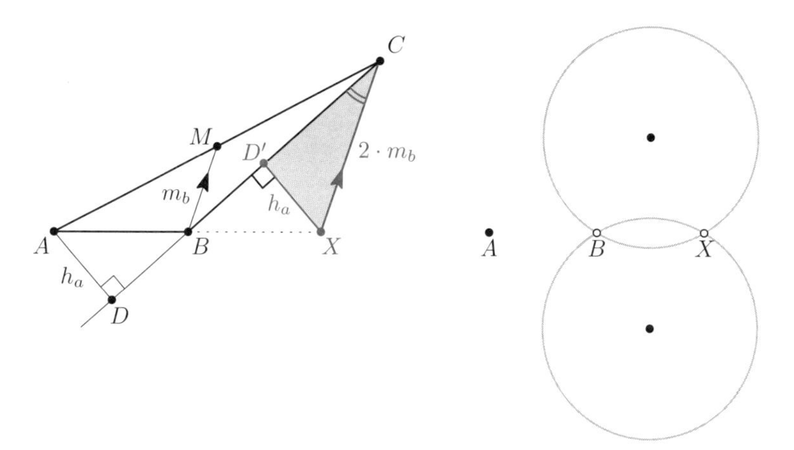

On the other hand, if for some point C the measure of $\angle BCX$ is 30° or 150° then by reverse chain of thoughts we get $h_a = m_b$.

Therefore the sought locus is the union of all the points on the two circles passing through B and X whose centers form equilateral triangles with the points B and X, but the points B and X themselves.

32. [MEMO 2011, Michal Rolínek and Josef Tkadlec] Let ABC be an acute triangle with altitudes BB_0 and CC_0. Point P is given such that the line PB is tangent to the circumcircle of triangle PAC_0 and the line PC is tangent to the circumcircle of triangle PAB_0. Prove that AP is perpendicular to BC.

First Proof. Lines AP and BC are perpendicular if and only if $AB^2 + CP^2 = AC^2 + BP^2$ (see Proposition 1.22). Using Power of a Point for B, C with respect to the circumcircles of triangles PAC_0, PAB_0, respectively, we get rid of point P and obtain

$$AB^2 + CP^2 = AB^2 + CA \cdot CB_0 = c^2 + b \cdot (a\cos\angle C),$$
$$AC^2 + PB^2 = AC^2 + BA \cdot BC_0 = b^2 + c \cdot (a\cos\angle B),$$

Finally, the Law of Cosines yields

$$ba\cos\angle C = \frac{1}{2}(a^2 + b^2 - c^2) \quad \text{and} \quad ca\cos\angle B = \frac{1}{2}(a^2 + c^2 - b^2)$$

implying that both right-hand sides of the former equalities are equal to $\frac{1}{2}(a^2 + b^2 + c^2)$.

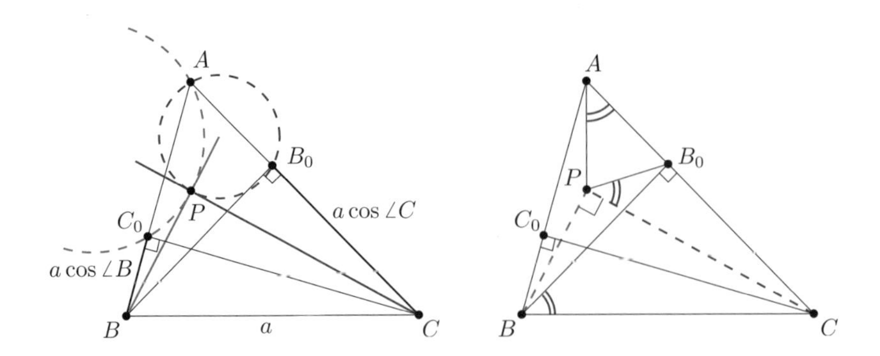

Second Proof. Using Power of a Point as in the first proof, we obtain $BP^2 = BA \cdot BC_0 = ac\cos\angle B$ and $CP^2 = CA \cdot CB_0 = ab\cos\angle C$. Hence

$$BP^2 + CP^2 = a(c\cos\angle B + b\cos\angle C) = a^2,$$

where we have recognized $c\cos\angle B + b\cos\angle C$ as the length of BC split at the foot of the altitude from A. Hence BPC is a right triangle and BCB_0PC_0 is a cyclic pentagon. Finally, from tangency (see Proposition 1.34) and concyclicity we obtain

$$\angle PAC = \angle CPB_0 = \angle CBB_0 = 90^\circ - \angle C.$$

Hence AP is perpendicular to BC.

33. [Moscow Math Olympiad 2011] Let ABC be a triangle. Point O in its interior satisfies $\angle OBA = \angle OAC$, $\angle BAO = \angle OCB$, and $\angle BOC = 90^\circ$. Find AC/OC.

Solution. It is somewhat difficult to draw an accurate diagram for this problem since for a fixed triangle ABC there does not necessarily have to exist point O in its interior satisfying all the three requirements. Hence we start with right triangle BOC and aim to construct point A such that the remaining two equalities hold too.

Since $\angle BAO = \angle OCB$, the segment OB is visible from A and C under the same angle and consequently A has to lie on the arc $BC'O$ where C' is the reflection of C about BO. The equality $\angle OBA = \angle OAC$ then implies that CA is tangent to the circumcircle of $BOAC'$ (see Proposition 1.34), so our diagram is complete.

Now it suffices to combine Power of a Point and symmetry about BO to obtain

$$CA^2 = CO \cdot CC' = CO \cdot (2 \cdot CO) = 2 \cdot CO^2.$$

Therefore the answer is $AC/OC = \sqrt{2}$.

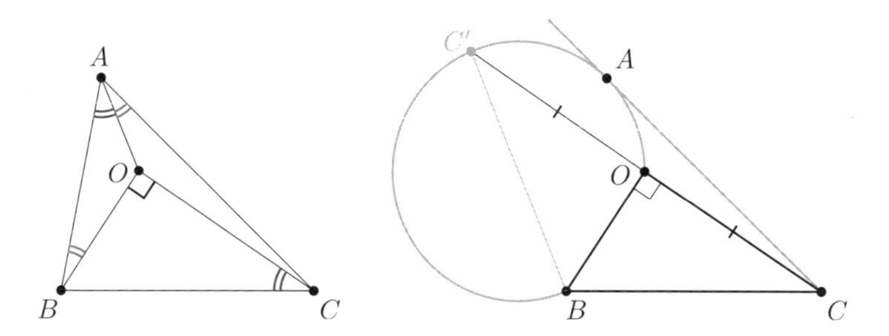

34. [Poland 2007] Let $ABCD$ be a cyclic quadrilateral ($AB \neq CD$). Quadrilaterals $AKDL$ and $CMBN$ are rhombi with equal sides. Prove that points K, L, M, N lie on a single circle.

    Proof. Given a cyclic quadrilateral $ABCD$, how do we find points K and L such that $AKDL$ is a rhombus with given side length d? Of course, we have to intersect circles ω_a, ω_d with centers A, D, respectively, and common radius d. Let us view points K, L (and likewise M, N) from this perspective instead.

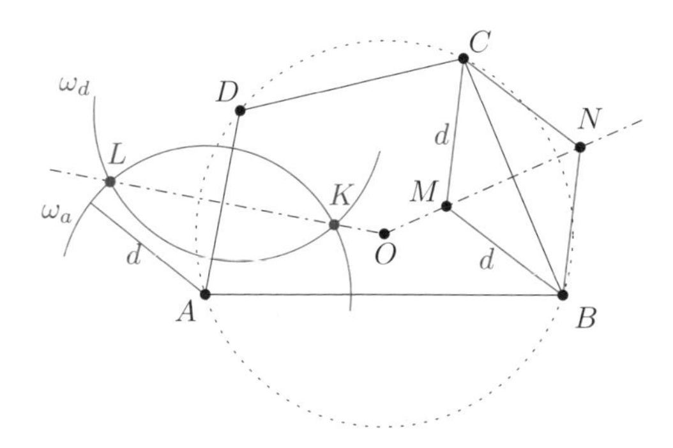

    As KL is the perpendicular bisector of AD, it passes through the circumcenter O of $ABCD$ and the same holds for MN, the bisector of BC. Hence it is reasonable to employ Power of a Point and try to prove $\overline{OK} \cdot \overline{OL} = \overline{OM} \cdot \overline{ON}$.

    This is quickly carried out once we observe

$$\overline{OK} \cdot \overline{OL} = p(O, \omega_a) = OA^2 - d^2,$$
$$\overline{OM} \cdot \overline{ON} = p(O, \omega_b) = OB^2 - d^2,$$

    where we invoked the very definition of Power of a Point. Indeed, as $OA = OB$, the result follows.

35. Let ABC be a triangle with inradius r and let ω be a circle of radius $a < r$ inscribed in angle BAC. Tangents from B and C to ω (different

from the triangle sides) intersect at point X. Show that the incircle of triangle BCX is tangent to the incircle of triangle ABC.

Proof. Denote the points of contact of the incircle of triangle BCX with its sides BC, CX, XB by D, E, F, respectively. Since both the incircles of triangle ABC and triangle BCX are tangent to BC, we aim to prove that they are tangent to it at the same point. For this it suffices to prove $BD - DC = \frac{1}{2}(a - b + c) - \frac{1}{2}(a + b - c) = AB - AC$ (see Proposition 1.15).

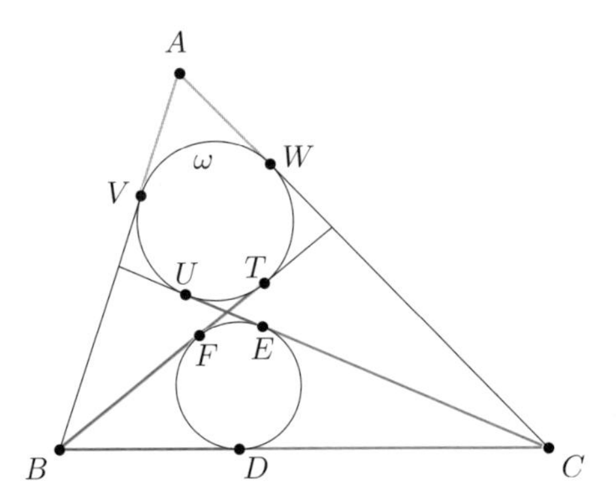

Let BX, CX, BA, CA be tangent to ω at T, U, V, W, respectively. Then $FT = UE$ (symmetry) and by Equal Tangents,

$$BD - DC = BF - EC = BT - UC = BV - WC = AB - AC,$$

which finishes the proof.

36. [IMO 2003] Let $ABCD$ be a cyclic quadrilateral. Let P, Q, R be the feet of the perpendiculars from D to the lines BC, CA, AB, respectively. Show that $PQ = QR$ if and only if the bisectors of $\angle ABC$ and $\angle ADC$ are concurrent with AC.

Proof. First, we explore the concurrence of the angle bisectors on AC as it happens if and only if both angle bisectors divide AC in the same ratio. If we use the Angle Bisector Theorem in triangles ABC and ADC, we can restate this equivalently as

$$\frac{AD}{CD} = \frac{AB}{BC}.$$

Let's now focus on the second statement. Note that points D, Q, P, C lie on a circle with diameter DC and likewise points D, Q, A, R lie on a circle with diameter AD. This enables us to find the lengths of the

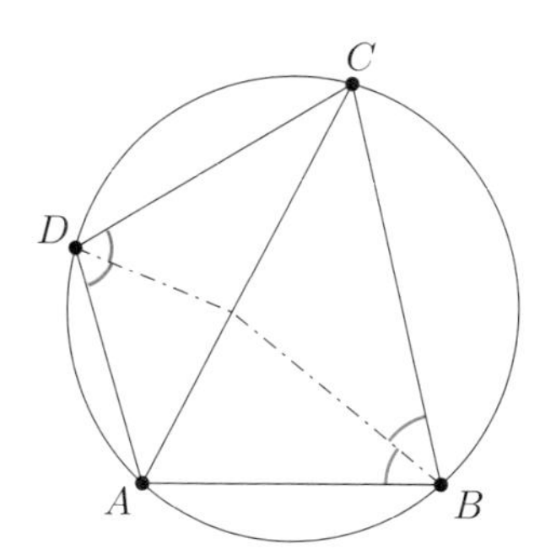

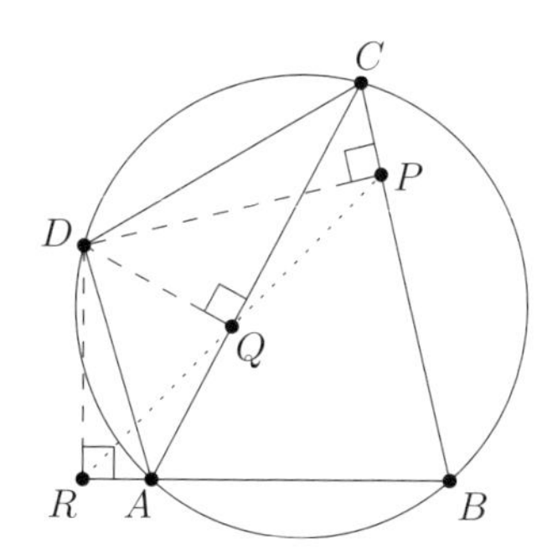

chords PQ and QR from the Extended Law of Sines as

$$\frac{PQ}{\sin\angle QCP} = CD, \quad \frac{QR}{\sin\angle QAR} = AD.$$

Now since $\sin\angle QCP = \sin\angle ACB$ and $\sin\angle QAR = \sin\angle BAC$, we can say that $PQ = QR$ if and only if

$$\frac{AD}{CD} = \frac{\sin\angle ACB}{\sin\angle BAC} = \frac{AB}{BC},$$

where the last equality is just the Law of Sines in triangle ABC. We have shown that both statements are equivalent to the same metric condition imposed on $ABCD$. We are done.

Remark. Note that throughout the whole proof we did not need the fact that $ABCD$ was cyclic.

37. Let X be a point on the circumcircle of a cyclic quadrilateral $ABCD$. Denote by E, F, G, and H the projections of X onto lines AB, BC, CD, DA, respectively. Prove that

$$BE \cdot CF \cdot DG \cdot AH = AE \cdot BF \cdot CG \cdot DH.$$

Proof. We drop one more perpendicular from X, this time to the line BD, and denote its foot by Z. We recall the Simson line (see Example 1.15) and deduce that triads of points E, Z, H and Z, G, F are collinear.

In order to produce such a huge equality, we use twice Menelaus' Theorem. Once for triangle BAD and points E, Z, H and the second time for triangle BCD and points Z, G, F. We obtain

$$\frac{BZ}{DZ} \cdot \frac{DH}{AH} \cdot \frac{AE}{BE} = 1 \quad \text{and} \quad \frac{BZ}{DZ} \cdot \frac{DG}{CG} \cdot \frac{CF}{BF} = 1.$$

Expressing the ratio BZ/DZ from both equations yields

$$\frac{AH}{DH} \cdot \frac{BE}{AE} = \frac{BZ}{DZ} = \frac{CG}{DG} \cdot \frac{BF}{CF},$$

which after expanding is exactly what we wanted to prove.

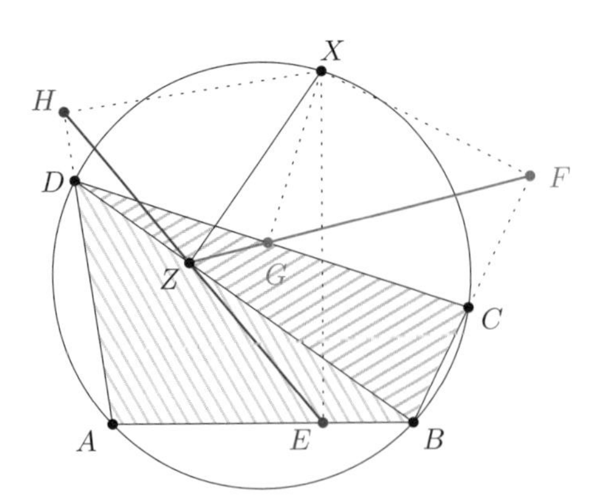

38. Newton-Gauss[2] line.

Let $ABCD$ be a convex quadrilateral. Denote by Q the intersection of AD and BC and by R the intersection of AB and CD. Let X, Y, and Z be the midpoints of AC, BD, and QR, respectively. Prove that X, Y, and Z lie on a single line.

Proof. Proving collinearity when midpoints are involved should invoke using Menelaus' Theorem. But for which triangle? We add more midpoints to find one and as usual, we expect to produce midlines. One of the possible choices are the midpoints of the sides of triangle ABQ.

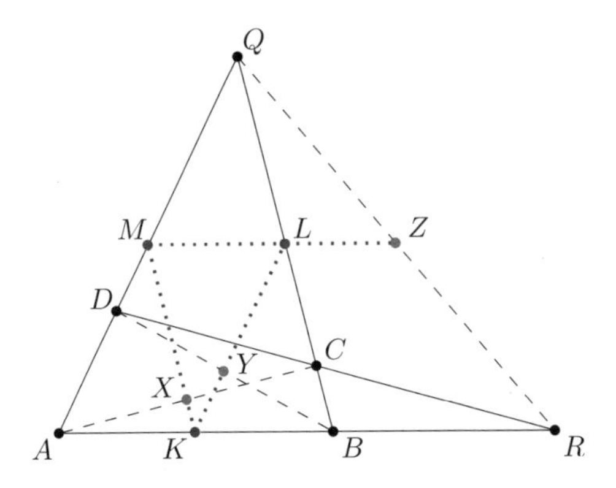

We claim that if we denote by K, L, M the midpoints of AB, BQ, and QA, then the points X, Y, and Z lie on the sides of triangle KLM (possibly extended). Indeed, as ML and LZ are midlines in triangles QAB, QBR, respectively, they are both parallel to AB and thus points M, L, and Z are collinear. The collinearity of M, X, K and K, Y, L is proved analogously.

Now by Menelaus' Theorem in triangle KLM for points X, Y, and Z

[2]Johann Carl Friedrich Gauss (1777–1855) was a German mathematician and physicist.

we need to prove that

$$\frac{MX}{XK} \cdot \frac{KY}{YL} \cdot \frac{LZ}{ZM} = 1.$$

But recalling that the length of a midline is half the length of the corresponding side, these ratios may be "projected" on the sides of triangle ABQ. We have

$$\frac{MX}{XK} = \frac{QC}{CB}, \quad \frac{KY}{YL} = \frac{AD}{DQ}, \quad \frac{LZ}{ZM} = \frac{BR}{RA}.$$

This enables us to forget all the midpoints, since now it suffices to prove an identity concerning only points A, B, C, D, P, and Q. Namely

$$\frac{QC}{CB} \cdot \frac{BR}{RA} \cdot \frac{AD}{DQ} = 1.$$

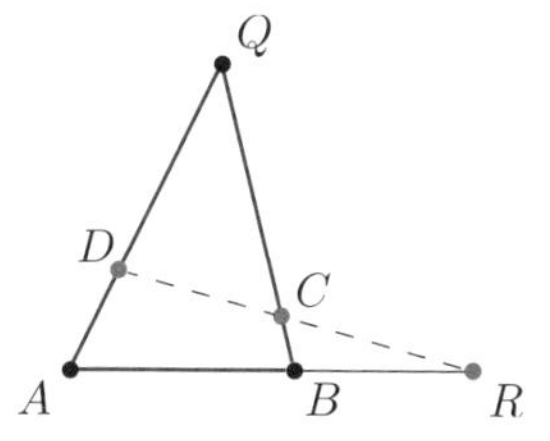

We may celebrate, as the latter is true by Menelaus' Theorem applied to triangle ABQ and collinear points D, C, and R.

39. [Moscow Math Olympiad 2009] In acute triangle ABC let A_1, B_1 be the points of tangency of A-excircle with BC and B-excircle with AC, respectively. Let H_1, H_2 be the orthocenters of triangles CAA_1 and CBB_1, respectively. Prove that H_1H_2 is perpendicular to the angle bisector of $\angle ACB$.

Proof. Let us view the orthocenter H_1 as the intersection of altitudes AA' and A_1A_1' and the orthocenter H_2 as the intersection of altitudes BB' and B_1B_1'. A reasonable way to make use of the fact that A_1, B_1 are the points of contact of the excircles is to recall $AB_1 = \frac{1}{2}(a+b-c) = BA_1$ (see Proposition 1.15(c)). In fact, we will use only $AB_1 = BA_1$.

Note that the angle $\angle C$ is contained by segments AB_1 and $A'B_1'$ and also by BA_1 and $B'A_1'$. Thus, projecting the equal segments AB_1, BA_1 onto the lines BC, AC we infer

$$A'B_1' = AB_1 \cos\angle C = BA_1 \cos\angle C = B'A_1'.$$

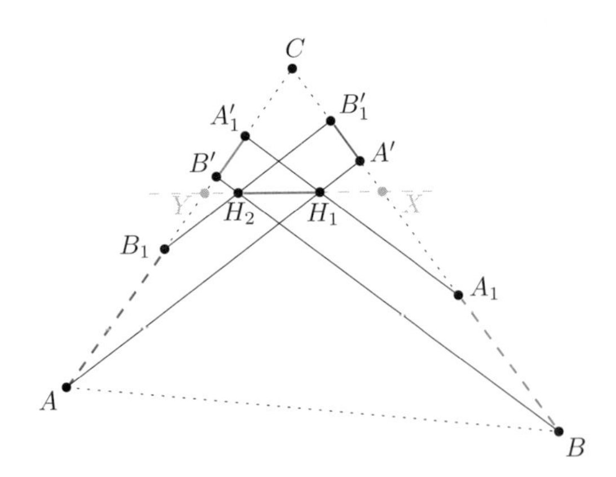

Now focus on the segment H_1H_2. We have just shown that its projections to the lines AC, BC are of the same length. The result should be apparent. Placing the bisector of $\angle BCA$ vertically, it is quite plausible that if the segment H_1H_2 was skew, one of its projections would be longer than the second one. Let us prove it rigorously.

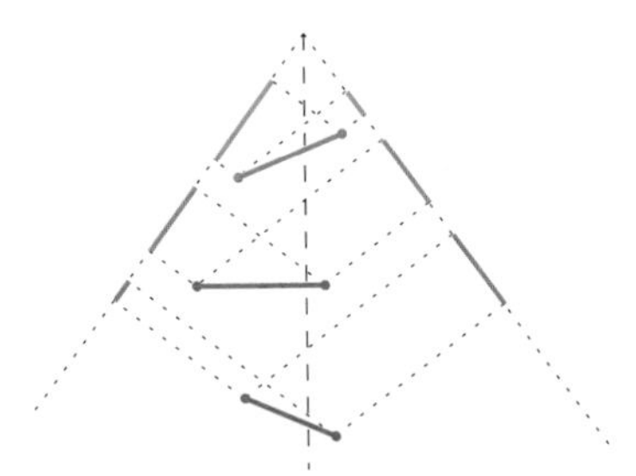

Since $\angle C$ is acute and the points C, B_1, A and C, A_1, B lie on the sides of $\angle C$ in this order, line H_1H_2 intersects the segments BC, AC. Denote the intersections by X, Y, respectively.

Similarly as before, we obtain

$$\cos \angle CXY = \frac{A'B'_1}{H_1H_2} = \frac{B'A'_1}{H_1H_2} = \cos \angle XYC.$$

Thus, $\angle CXY = \angle XYC$ and the triangle CXY is isosceles. Its base is then indeed perpendicular to the angle bisector of $\angle BCA$.

40. [China 1997] A circle ω with center O is internally tangent to two circles in its interior at points S and T which are not diametrically opposite. Suppose the two circles intersect at M and N with N closer to ST. Show that $OM \perp MN$ if and only if S, N, T are collinear.

    Proof. To make use of the tangency, draw common tangents at points S and T and denote by X their intersection. Since $XS = XT$, point X

has equal power to the two small circles and thus it lies on their radical axis MN.

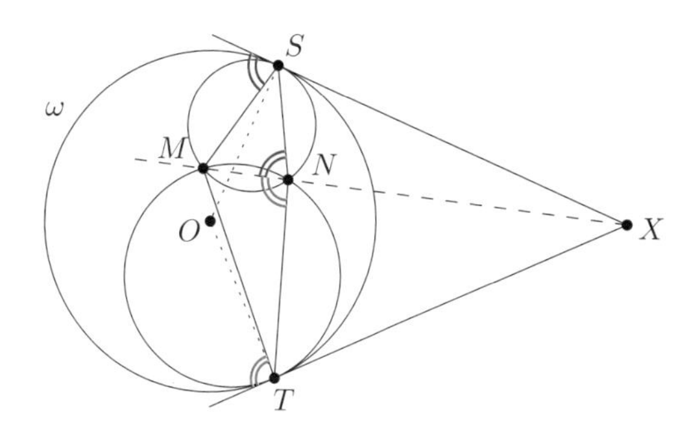

The condition $OM \perp MN$ is equivalent to $\angle OMX = 90°$. As points S and T both lie on a circle with diameter OX, the latter happens if and only if the points S, M, T and X are concyclic.

On the other hand, we have $\angle SNM = 180° - \angle MSX$ and $\angle MNT = 180° - \angle XTM$ (see Proposition 1.34). Summing these two relations we deduce that the points S, N, T are collinear if and only if $SMTX$ is cyclic and thus we are done.

41. Let $ABCD$ be a quadrilateral with an inscribed circle ω and let the points of tangency of the incircle with sides AB, BC, CD, DA be K, L, M, N, respectively. Prove that the lines AC, BD, KM, and LN are concurrent.

Proof. We start by observing that $\angle DMK = \angle MKA$, as both angles correspond to the same arc MK of ω (see Proposition 1.34). Our strategy will be ratios and the Law of Sines.

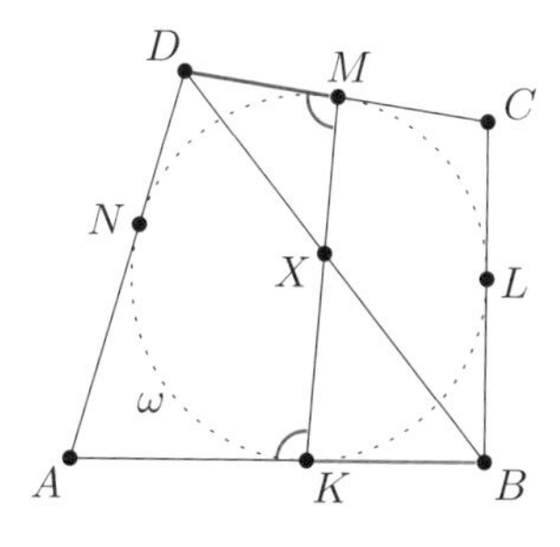

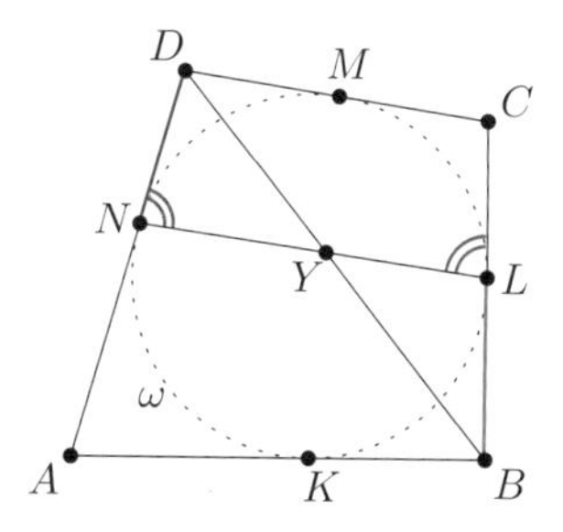

Let KM intersect BD at X. We intend to find the ratio in which X divides BD. The Law of Sines in triangles XMD and KBX yields

$$DX = MD \cdot \frac{\sin \angle DMX}{\sin \angle MXD}, \quad XB = KB \cdot \frac{\sin \angle BKX}{\sin \angle KXB}.$$

Since $\sin \angle DMX = \sin \angle BKX$, we can find the sought-after ratio as

$$\frac{DX}{XB} = \frac{MD}{KB}.$$

Next, we intersect BD and NL at point Y and find how it divides BD. After analogous calculation, we obtain

$$\frac{DY}{YB} = \frac{ND}{LB},$$

and we observe that due to Equal Tangents from B and D, the right-hand sides of the last two relations are equal and thus $Y = X$ and BD, KM, and LN are indeed concurrent.

In the same fashion, we show that also AC, KM, and LN are concurrent.

42. Orthologic triangles.

Let ABC and $A'B'C'$ be two triangles in plane. Show that the perpendiculars from A' to BC, from B' to CA and from C' to AB (denote their feet by X, Y, and Z, respectively) are concurrent if and only if the perpendiculars from A to $B'C'$, from B to $C'A'$, and from C to $A'B'$ are concurrent.

Proof. By Carnot's Theorem (see Introductory Problem 49) the perpendiculars to the sides of triangle ABC are concurrent if and only if

$$BX^2 + CY^2 + AZ^2 = CX^2 + AY^2 + BZ^2.$$

Next, we rewrite this condition so that points A', B', and C' are involved.

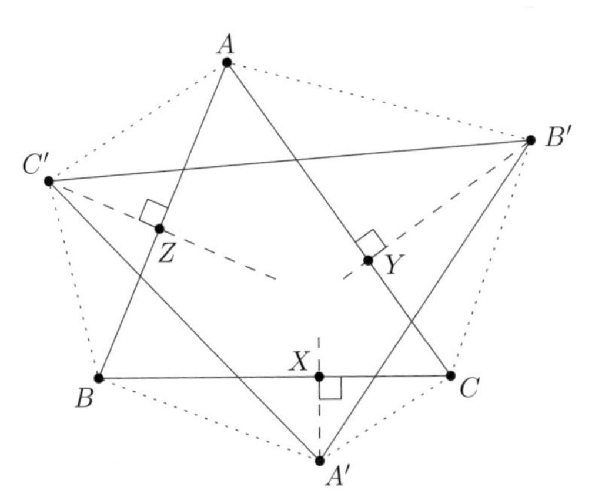

From the perpendicularity criterion (see Proposition 1.22) for $A'X \perp BC$ we learn that

$$BX^2 - CX^2 = A'B^2 - A'C^2.$$

Similarly, we obtain

$$CY^2 - AY^2 = B'C^2 - B'A^2 \quad \text{and} \quad AZ^2 - BZ^2 = C'A^2 - C'B^2.$$

Thus the original condition can be equivalently rewritten as

$$A'B^2 + B'C^2 + C'A^2 = B'A^2 + C'B^2 + A'C^2.$$

It remains to realize that the concurrence of the perpendiculars to the sides of triangle $A'B'C'$ can be treated in the same way and turns out to be equivalent to the same condition, from which the conclusion follows.

43. [All-Russian Olympiad 1994] Let ABC be a triangle with medians m_a, m_b, m_c and circumradius R. Prove that

$$\frac{b^2+c^2}{m_a} + \frac{c^2+a^2}{m_b} + \frac{a^2+b^2}{m_c} \le 12R.$$

Proof. We avoid tough computations by geometric argument but it's going to be tricky! We divide the equality by 2 and recall the median formula (see Corollary 1.24). Then we rewrite each term as follows

$$\frac{b^2+c^2}{2m_a} = \frac{\frac{1}{2}(b^2+c^2) - \frac{a^2}{4}}{m_a} + \frac{a^2}{4m_a} = m_a + \frac{a^2}{4m_a}.$$

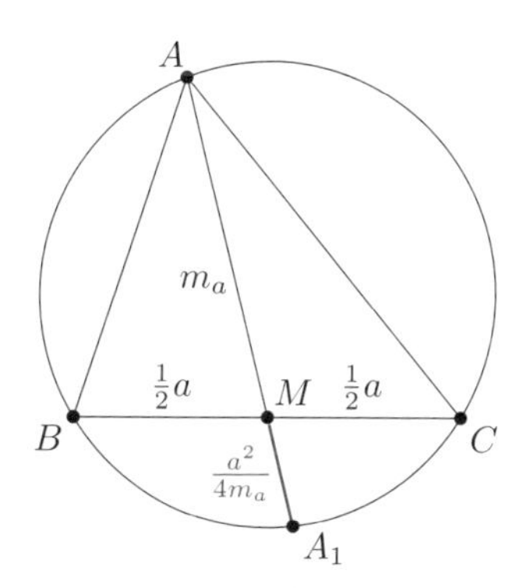

The clever idea now is to apply Power of a Point. Indeed, denote by M the midpoint of BC and by A_1 the point where the extended median meets the circumcircle ω for the second time. Then taking the power of M with respect to ω gives

$$\frac{a^2}{4} = MB \cdot MC = m_a \cdot MA_1, \quad \text{hence} \quad MA_1 = \frac{a^2}{4m_a}.$$

So in fact

$$\frac{b^2+c^2}{2m_a} = m_a + MA_1 = AA_1 \le 2R,$$

where the last inequality holds because diameter is the longest chord. Adding similar inequalities for the other two fractions gives the result.

44. [Paul Erdős] Show that in acute triangle ABC we have $r+R \leq h$, where r, R, and h are the inradius, circumradius and the longest altitude, respectively.

Proof. Denote by I the incenter of triangle ABC and by D, E, F the points of contact of the incircle with the sides BC, CA, AB, respectively. Segments ID, IE, IF split triangle ABC into three quadrilaterals. Triangle ABC is acute, so its circumcenter O lies inside of

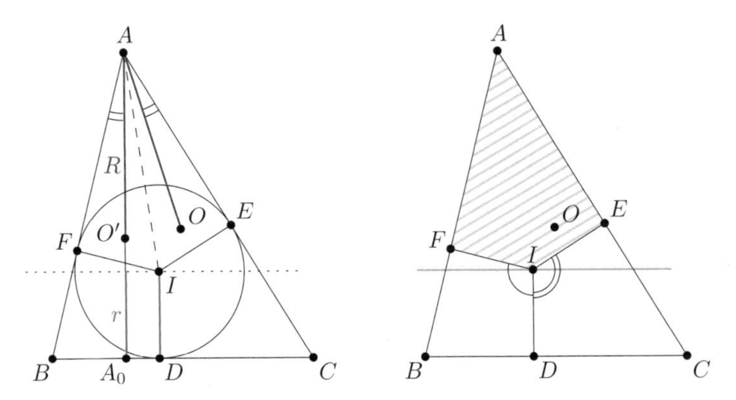

it. Without loss of generality suppose it lies inside $AFIE$ (including the boundary), place BC horizontally with A on top and drop altitude AA_0.

It suffices to prove that $r+R \leq AA_0$. As O lies inside the quadrilateral $AFIE$ which is symmetric about AI, so does its reflection O' about AI. Furthermore, since "H and O are friends" (see Proposition 1.47), point O' lies on the altitude AA_0.

What remains to prove now is $A_0O' \geq DI$. But that's obvious! Since both $\angle B$ and $\angle C$ are acute, both $\angle DIE$ and $\angle DIF$ are obtuse implying that point I is "the lowest" point of quadrilateral $AFIE$. Point O' (being inside $AFIE$) is thus not below I and we may conclude.

45. [USAMO 2012, Titu Andreescu and Cosmin Pohoaţă] Let P be a point in the plane of triangle ABC, and ℓ a line passing through P. Let A', B', C' be the points where the reflections of lines PA, PB, PC with respect to ℓ intersect lines BC, AC, AB respectively. Prove that A', B', C' are collinear.

Proof. First we note that if any of the points A', B', C' coincides with one of the vertices, say $A' = C$, then lines AP and CP are symmetric with respect to ℓ and thus also $C' = A$, which means A', B', C' all lie on AC.

For the general case we intend to use Menelaus' Theorem. We begin with the ratio $BA'/A'C$ which we find from the Ratio Lemma (see Proposition 1.18) as

$$\frac{BA'}{A'C} = \frac{BP}{CP} \cdot \frac{\sin \angle BPA'}{\sin \angle A'PC}.$$

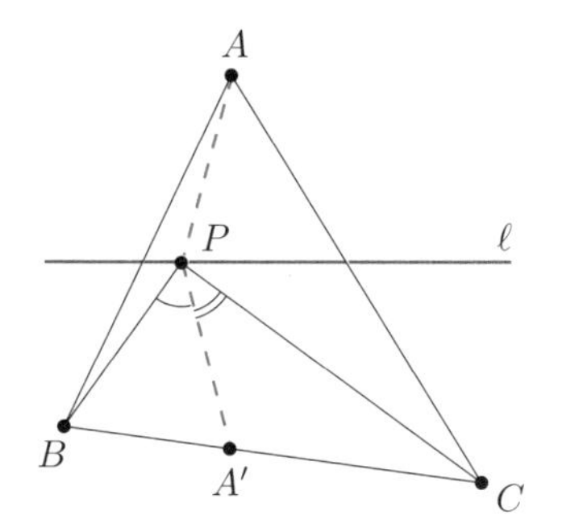

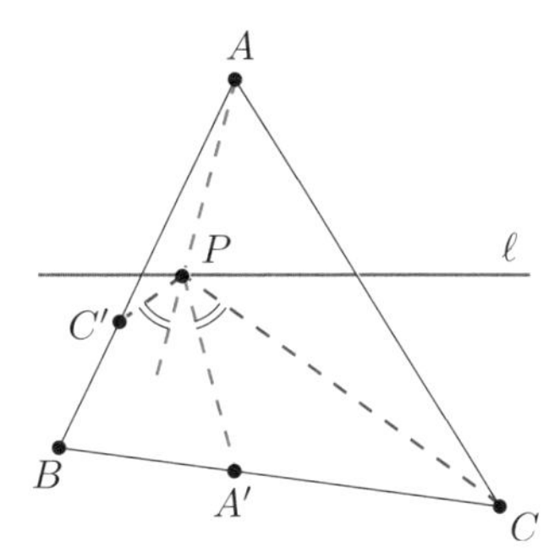

Now the crucial observation is that $\angle(C'P, PA) = -\angle(CP, PA')$ since these angles correspond in symmetry with respect to ℓ. Thus,

$$\sin \angle APC' = \sin \angle A'PC$$

and we may rewrite the Ratio Lemma as

$$\frac{BA'}{A'C} = \frac{BP}{CP} \cdot \frac{\sin \angle BPA'}{\sin \angle APC'}$$

From here it is not hard to see that once we express the ratios $CB'/B'A$ and $AC'/C'B$ in the same manner and plug it in Menelaus' Theorem, everything cancels out. Indeed, we have

$$\begin{aligned}
&\frac{BA'}{A'C} \cdot \frac{CB'}{B'A} \cdot \frac{AC'}{C'B} = \\
&\quad = \frac{BP}{CP} \cdot \frac{\sin \angle BPA'}{\sin \angle APC'} \cdot \frac{CP}{AP} \cdot \frac{\sin \angle CPB'}{\sin \angle BPA'} \cdot \frac{AP}{BP} \cdot \frac{\sin \angle APC'}{\sin \angle CPB'} \\
&\quad = 1.
\end{aligned}$$

One last thing we need to take care of is, since we used the undirected version of Menelaus' Theorem, to verify that either 0 or 2 points A', B', C' lie on the perimeter of triangle ABC. For this we use directed angles. If we denote by α, β and γ the angles $\angle(AP, \ell)$, $\angle(BP, \ell)$, $\angle(CP, \ell)$, respectively, we see that PA' lies between BP and CP (i.e. A' lies on the segment BC) if and only if

$$\beta > 180^\circ - \alpha > \gamma, \quad \text{or equivalently} \quad \beta + \alpha > 180^\circ > \gamma + \alpha,$$

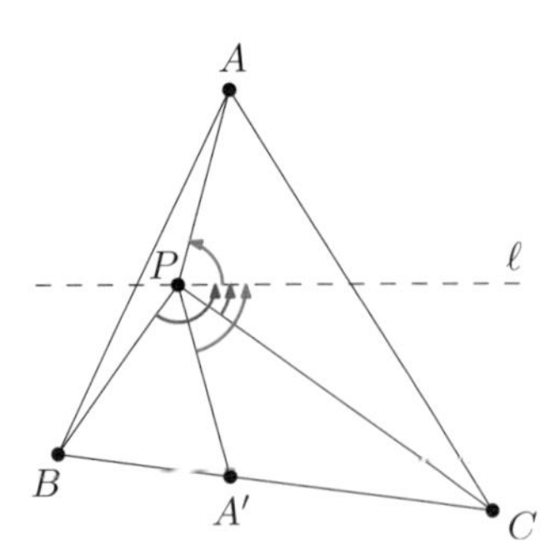

where the angles are no longer considered mod 180°. Analogous conditions hold for points B' and C' and it is easy to see that exactly 0 or 2 of them may be satisfied.

46. Let BC be the longest side of a scalene triangle ABC. Point K on the ray CA satisfies $KC = BC$. Similarly, point L on the ray BA satisfies $BL = BC$. Prove that KL is perpendicular to OI where O, I denote the circumcenter and the incenter of triangle ABC, respectively.

Proof. Denote the circumcircle of triangle ABC by Ω and its incircle by ω. Since the locus of points X, for which the difference $p(X,\Omega)-p(X,\omega)$ is equal to given constant, is a line perpendicular to OI (see Introductory Problem 53), it suffices to prove

$$p(K,\Omega) - p(K,\omega) = p(L,\Omega) - p(L,\omega).$$

Denoting the points of contact of the incircle with the sides BC, CA, AB by D, E, F, respectively, all the powers can be expressed easily in the xyz notation, which reduces the whole problem to some straightforward algebra. We ease our lives a bit by noting $KE = BD = y$ (symmetry about the angle bisector of $\angle C$) and compute

$$\begin{aligned} p(K,\Omega) - p(K,\omega) &= KA \cdot KC - KE^2 = (y-x)(y+z) - y^2 = \\ &= yz - x(y+z). \end{aligned}$$

Since the last expression is symmetric with respect to y and z, the proof is complete.

47. [USAMO 1991] Let D be an arbitrary point on the side BC of a given triangle ABC and let E be the intersection of AD and the second external common tangent of the incircles of triangles ABD and ACD. As D assumes all positions between B and C, prove that the point E traces an arc of a circle.

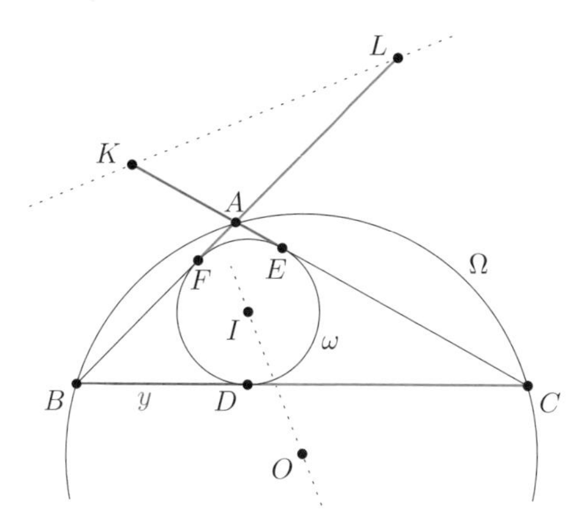

Proof. First, when D is approaching C, it seems plausible that E is becoming the point of contact of the incircle of triangle ABC with the side AC. Likewise, when D tends to B, it appears that E tends to the point of contact of the incircle of triangle ABC with AB. Hence it is natural to expect that the desired locus is the arc of the circle with center A and radius $x = \frac{1}{2}(b + c - a)$ (see Proposition 1.15) which lies inside the triangle ABC. Once we guessed it, it suffices to find AE in terms of the side lengths a, b, c only (it should turn out to be independent on the choice of D).

Denote the points of contact of the incircles of triangles ABD, ACD with the side BC by T, U, respectively, the distances from D to the vertices by $DA = d$, $DB = m$, $DC = n$, and finally recall that $DE = TU$ (see Proposition 1.13 (b)).

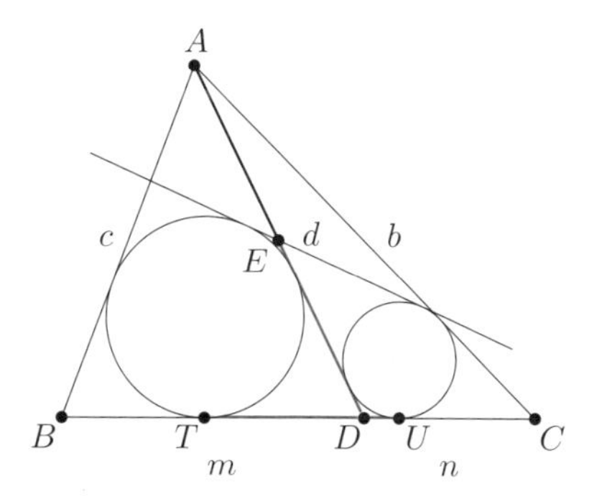

Hence

$$AE = d - ED = d - TU = d - (DT + DU).$$

As $DT = \frac{1}{2}(d + m - c)$ and $DU = \frac{1}{2}(d + n - b)$, we have

$$AE = \frac{1}{2}(b + c - m - n) = \frac{1}{2}(b + c - a)$$

as desired.

48. [IMO 2009] Let ABC be a triangle with circumcenter O. The points P and Q are interior points of the sides CA and AB, respectively. Let K, L and M be the midpoints of the segments BP, CQ and PQ, respectively, and let Γ be the circle passing through K, L, and M. Suppose that the line PQ is tangent to the circle Γ. Prove that $OP = OQ$.

Proof. Without even drawing the circle Γ, we translate the tangency as $\angle MLK = \angle QMK$ and $\angle LKM = \angle LMP$ (see Proposition 1.34).

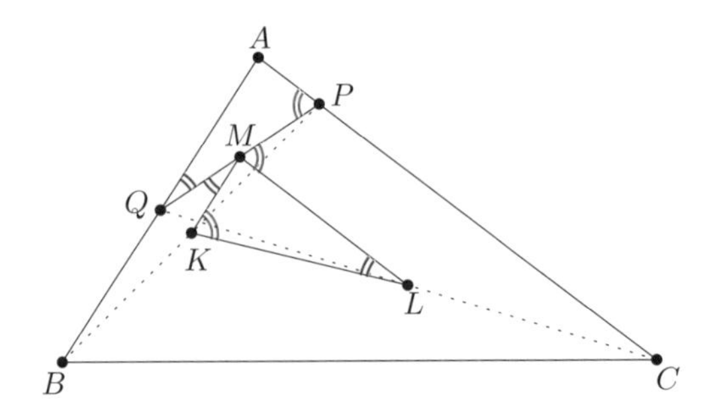

Next, we note that MK is the midline in triangle BPQ and ML is the midline in triangle CQP. Hence we can chase angles a little more

$$\angle MLK = \angle QMK = \angle MQA,$$

and similarly we obtain $\angle LKM = \angle APQ$ and thus triangles AQP and MLK are similar (AA). From the ratios we learn that

$$\frac{AP}{AQ} = \frac{MK}{ML} = \frac{\frac{1}{2}QB}{\frac{1}{2}PC},$$

which rewrites as $AP \cdot PC = AQ \cdot QB$. But this implies that P and Q have the same power with respect to the circumcircle of triangle ABC, therefore they have the same distance from its center O (see Proposition 1.40(a)). We are done.

49. [IMO 1995 shortlist] Let ABC be a non-right triangle. A circle ω passing through B and C intersects the sides AB and AC again at C' and B', respectively. Prove that BB', CC' and HH' are concurrent, where H and H' are the orthocenters of triangles ABC and $AB'C'$, respectively.

First Proof. We recall Introductory Problem 51 and apply it for triangle ABC with orthocenter H and cevians BB' and CC' and also for triangle $AB'C'$ with orthocenter H' and cevians $B'B$ and $C'C$. We learn that the line HH' is in fact the radical axis of circles with diameters BB' and CC' (call them ω_b and ω_c).

Moreover, we observe that BB' is precisely the radical axis of ω and ω_b, and CC' is the radical axis of ω and ω_c. Since pairwise radical axes of three circles intersect at their radical center (see Proposition 1.42), we are done.

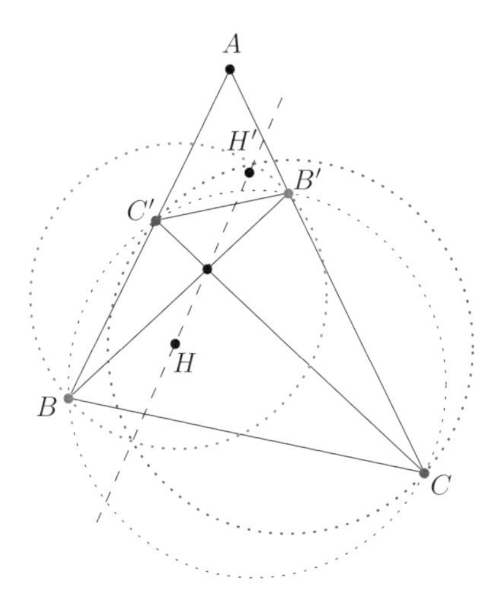

Second Proof. Let $P = BB' \cap CC'$ and also let $X = BH \cap CC'$ and $Y = CH \cap BB'$. Antiparallelism is the key ingredient in this solution.

Perpendiculars BH and CH to AC and AB are antiparallel in $\angle BAC$ and since $BCB'C'$ is cyclic, they are also antiparallel in $\angle BPC$ (see Corollary 1.36). Thus, in $\angle BPC$ both $B'C'$ and XY are antiparallel to BC implying that $XY \parallel B'C'$.

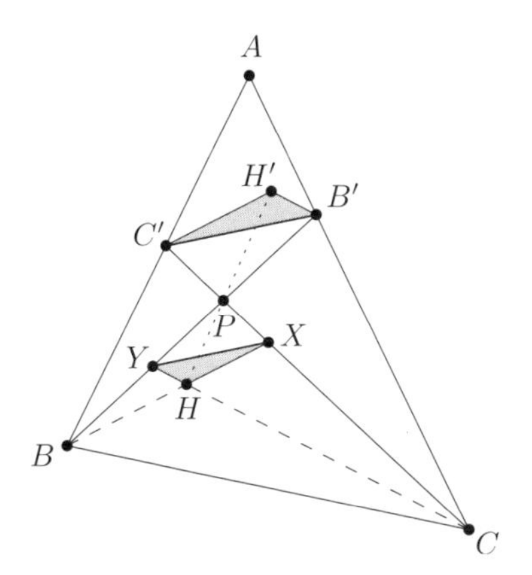

Furthermore, since $BH \parallel C'H'$ (perpendiculars to AC) and $CH \parallel B'H'$ (perpendiculars to AB) the triangles HXY and $H'C'B'$ have corresponding sides parallel and hence the quadrilaterals $PYHX$ and $PB'H'C$ are similar. Since the segments PH and PH' correspond, they are also parallel, thus HH' passes through P and we may conclude.

50. [USA TST 2000, Titu Andreescu] Let P be a point in the interior of triangle ABC with circumradius R. Prove that

$$\frac{AP}{a^2} + \frac{BP}{b^2} + \frac{CP}{c^2} \geq \frac{1}{R}.$$

Proof. Denote the projections of P onto BC, CA, AB by X, Y, Z, respectively, and recall the "key ingredient of the proof of the Erdős-Mordell inequality", namely the inequality

$$PA \sin \angle A \geq PY \sin \angle C + PZ \sin \angle B,$$

which compares the length of YZ to the length of its projection onto BC.

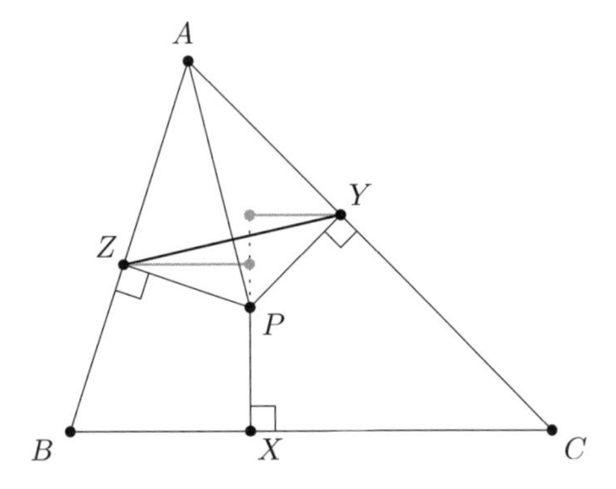

Expressing every sine from the Extended Law of Sines in triangle ABC and cancelling $2R$ this rewrites as

$$aPA \geq cPY + bPZ$$

or

$$\frac{PA}{a^2} \geq PY \cdot \frac{c}{a^3} + PZ \cdot \frac{b}{a^3}.$$

Likewise, we obtain

$$\frac{PB}{b^2} \geq PZ \cdot \frac{a}{b^3} + PZ \cdot \frac{c}{b^3},$$
$$\frac{PC}{c^2} \geq PX \cdot \frac{b}{c^3} + PY \cdot \frac{a}{c^3}.$$

Summing these inequalities, applying AM-GM inequality to the terms in parentheses, and finally recalling the area formula involving circumradius (see Proposition 1.25) we can estimate the left-hand side (LHS) of the given inequality as

$$LHS \geq PX\left(\frac{b}{c^3}+\frac{c}{b^3}\right)+PY\left(\frac{c}{a^3}+\frac{a}{c^3}\right)+PZ\left(\frac{a}{b^3}+\frac{b}{a^3}\right) \geq$$
$$\geq \frac{2\cdot PX}{bc}+\frac{2\cdot PY}{ca}+\frac{2\cdot PZ}{ab}=\frac{4[ABC]}{abc}=\frac{1}{R},$$

which is exactly what we wanted.

Equality in the first step requires YZ to be parallel to BC and so on. This occurs if and only if P is the circumcenter of triangle ABC. Equality in AM-GM requires $a=b=c$. Thus the equality holds if and only if triangle ABC is equilateral and P is its center.

51. [USAMO 2010, Titu Andreescu] Let $AXYZB$ be a convex pentagon inscribed in a semicircle of diameter AB. Denote by P, Q, R, S the feet of the perpendiculars from Y onto lines AX, BX, AZ, BZ, respectively. Prove that the acute angle formed by lines PQ and RS is half the size of $\angle ZOX$, where O is the midpoint of the segment AB.

    Proof. A line passing through two feet of perpendiculars should remind us of the Simson line (see Example 1.15).

    Denote by T the foot of perpendicular from Y onto line AB. Line PQ is the Simson line of point Y with respect to triangle ABX, hence it passes through T. Analogous reasoning shows that $T \in RS$.

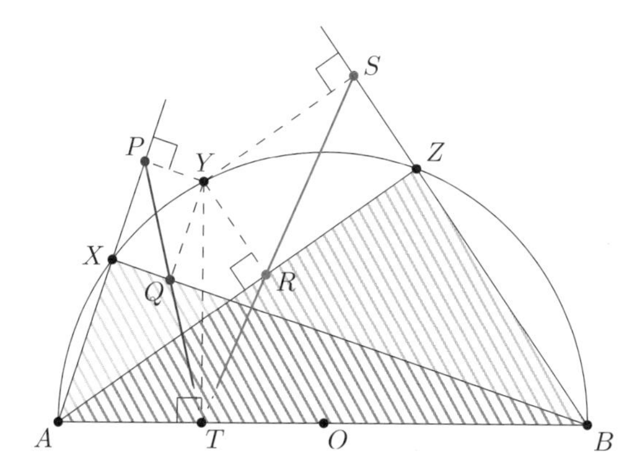

    On the other hand, $\angle ZOX$ is the central angle corresponding to the minor arc XZ. Its half is therefore simply the corresponding inscribed angle. Thus, it is enough to show that $\angle RTP = \angle RAP$ which is easily accomplished as points A, T, R, P all lie on a circle with diameter AY.

52. [Japan 2012] Let PAB and PCD be triangles such that $PA = PB$, $PC = PD$, and triads of points P, A, C and B, P, D are both collinear

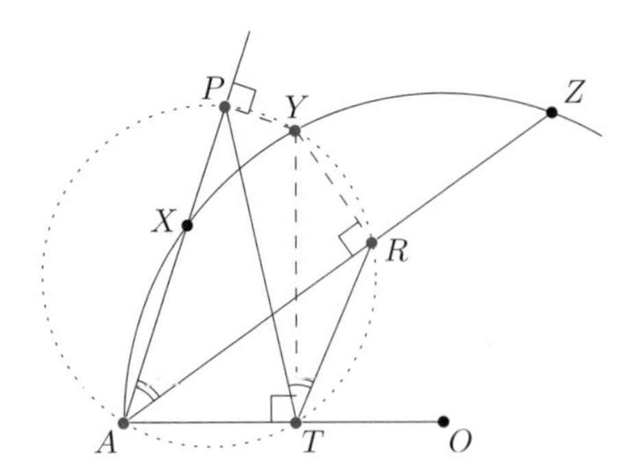

in this order. A circle ω_1 passing through A and C intersects a circle ω_2 passing through B and D at distinct points X, Y. Prove that the circumcenter of the triangle PXY is the midpoint of the segment formed by the centers O_1, O_2 of ω_1, ω_2, respectively.

Proof. Believe it or not, this is going to be a one-sentence proof:

Recalling that the locus of points with fixed sum of powers with respect to two given circles is a circle centered at the midpoint of the centers of the two circles (see Introductory Problem 53 (b)), it suffices to observe that

$$p(X, \omega_1) + p(X, \omega_2) = 0 + 0 = 0, \qquad p(Y, \omega_1) + p(Y, \omega_2) = 0 + 0 = 0,$$

and

$$p(P, \omega_1) + p(P, \omega_2) = PA \cdot PC + (-PB \cdot PD) = 0.$$

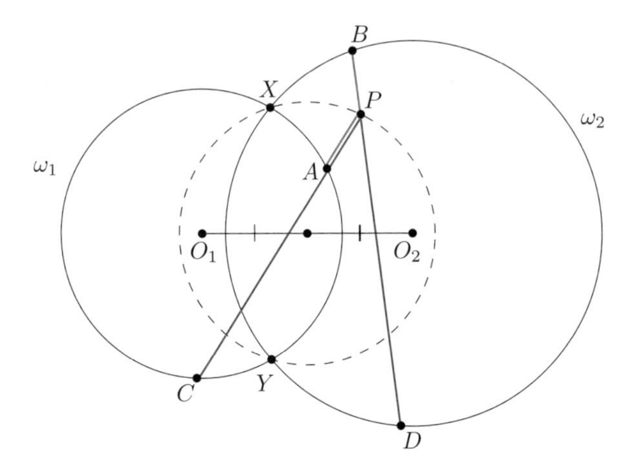

53. [IMO 2012, Josef Tkadlec] Let ABC be a triangle with $\angle BCA = 90°$, and let D be the foot of the altitude from C. Let X be a point in the interior of the segment CD. Let K be the point on the segment AX

such that $BK = BC$. Similarly, let L be the point on the segment BX such that $AL = AC$. Let M be the point of intersection of AL and BK.

Show that $MK = ML$.

Proof. Let ω_a and ω_b be the circles with centers A and B, respectively, passing through L and K, respectively. Since $BC = BK$ and $AC = AL$, both these circles pass through C and as the angle by C is right, ω_a is tangent to BC and ω_b to AC. Moreover, the radical axis of ω_a and ω_b is the line passing through C perpendicular to AB, i.e. the altitude CD.

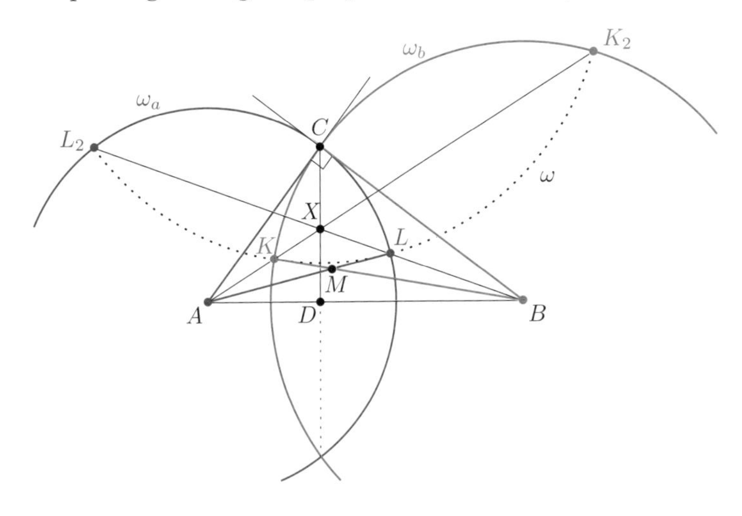

Hence if we let AX meet ω_b for the second time at K_2 and likewise BX meet ω_a again at L_2 then the Radical Lemma (see Proposition 1.43) implies that the quadrilateral L_2KLK_2 is cyclic. Denote its circumcircle by ω. Now the power of A with respect to ω_b gives

$$AK \cdot AK_2 = AC^2 = AL^2,$$

thus AL is tangent to ω (and so is ML). Analogously, MK is tangent to ω. Hence $MK = ML$ by Equal Tangents.

Further Reading

1. Altschiller-Court, N., *College Geometry, an Introduction to the Modern Geometry of the Triangle and the Circle*, Dover publications, 2007.
2. Andreescu, T.; Feng, Z., *101 Problems in Algebra from the Training of the USA IMO Team*, Australian Mathematics Trust, 2001.
3. Andreescu, T.; Feng, Z., *102 Combinatorial Problems from the Training of the USA IMO Team*, Birkhäuser, 2002.
4. Andreescu, T.; Feng, Z., *103 Trigonometry Problems from the Training of the USA IMO Team*, Birkhäuser, 2004.
5. Andreescu, T.; Andrica, D.; Feng, Z., *104 Number Theory Problems from the Training of the USA IMO Team*, Birkhäuser, 2006.
6. Andreescu, T.; Feng, Z.; Loh, P., *USA and International Mathematical Olympiads 2004*, Mathematical Association of America, 2005.
7. Andreescu, T.; Feng, Z., *USA and International Mathematical Olympiads 2003*, Mathematical Association of America, 2004.
8. Andreescu, T.; Feng, Z., *USA and International Mathematical Olympiads 2002*, Mathematical Association of America, 2003.
9. Andreescu, T.; Feng, Z., *USA and International Mathematical Olympiads 2001*, Mathematical Association of America, 2002.
10. Andreescu, T.; Feng, Z., *USA and International Mathematical Olympiads 2000*, Mathematical Association of America, 2001.
11. Andreescu, T.; Feng, Z.; Lee, G.; Loh, P., *Mathematical Olympiads: Problems and Solutions from around the World, 2001–2002*, Mathematical Association of America, 2004.
12. Andreescu, T.; Feng, Z.; Lee, G., *Mathematical Olympiads: Problems and Solutions from around the World, 2000–2001*, Mathematical Association of America, 2003.

13. Andreescu, T.; Feng, Z., *Mathematical Olympiads: Problems and Solutions from around the World, 1999–2000*, Mathematical Association of America, 2002.

14. Andreescu, T.; Feng, Z., *Mathematical Olympiads: Problems and Solutions from around the World, 1998–1999*, Mathematical Association of America, 2000.

15. Andreescu, T.; Kedlaya, K., *Mathematical Contests 1997–1998: Olympiad Problems from around the World, with Solutions*, American Mathematics Competitions, 1999.

16. Andreescu, T.; Kedlaya, K., *Mathematical Contests 1996–1997: Olympiad Problems from around the World, with Solutions*, American Mathematics Competitions, 1998.

17. Andreescu, T.; Kedlaya, K.; Zeitz, P., *Mathematical Contests 1995–1996: Olympiad Problems from around the World, with Solutions*, American Mathematics Competitions, 1997.

18. Andreescu, T.; Enescu, B., *Mathematical Olympiad Treasures*, 2nd edition, Birkhäuser, 2011.

19. Andreescu, T.; Gelca, R., *Mathematical Olympiad Challenges*, 2nd edition, Birkhäuser, 2009.

20. Andreescu, T.; Andrica, D.; Cucurezeanu, I., *An Introduction to Diophantine Equations: A Problem-Based Approach*, Birkhäuser, 2010.

21. Andreescu, T.; Andrica, D., *360 Problems for Mathematical Contests*, GIL Publishing House, 2003.

22. Andreescu, T.; Andrica, D., *Complex Numbers from A to Z*, Birkhäuser, 2004.

23. Andreescu, T.; Feng, Z., *A Path to Combinatorics for Undergraduate Students: Counting Strategies*, Birkhäuser, 2003.

24. Andreescu, T.; Andrica, D., *Number Theory - A Problem Solving Approach*, Birkhäuser, 2009.

25. Coxeter, H. S. M.; Greitzer, S. L., *Geometry Revisited*, New Mathematical Library, Vol. 19, Mathematical Association of America, 1967.

26. Coxeter, H. S. M., *Non-Euclidean Geometry*, The Mathematical Association of America, 1998.

27. Doob, M., *The Canadian Mathematical Olympiad 1969–1993*, University of Toronto Press, 1993.

28. Engel, A., *Problem-Solving Strategies*, Problem Books in Mathematics, Springer, 1998.

29. Fomin, D.; Kirichenko, A., *Leningrad Mathematical Olympiads 1987–1991*, MathPro Press, 1994.

30. Fomin, D.; Genkin, S.; Itenberg, I., *Mathematical Circles*, American Mathematical Society, 1996.

31. Gelca, R.; Andreescu, T., *Putnam and Beyond*, Springer, 2007.

32. Graham, R. L.; Knuth, D. E.; Patashnik, O., *Concrete Mathematics*, Addison-Wesley, 1989.

33. Greitzer, S. L., *International Mathematical Olympiads, 1959–1977*, New Mathematical Library, Vol. 27, Mathematical Association of America, 1978.

34. Holton, D., *Let's Solve Some Math Problems*, A Canadian Mathematics Competition Publication, 1993.

35. Kazarinoff, N. D., *Geometric Inequalities*, New Mathematical Library, Vol. 4, Random House, 1961.

36. Kedlaya, K; Poonen, B.; Vakil, R., *The William Lowell Putnam Mathematical Competition 1985–2000*, The Mathematical Association of America, 2002.

37. Klamkin, M., *International Mathematical Olympiads, 1978–1985*, New Mathematical Library, Vol. 31, Mathematical Association of America, 1986.

38. Klamkin, M., *USA Mathematical Olympiads, 1972–1986*, New Mathematical Library, Vol. 33, Mathematical Association of America, 1988.

39. Kürschák, J., *Hungarian Problem Book, volumes I & II*, New Mathematical Library, Vols. 11 & 12, Mathematical Association of America, 1967.

40. Kuczma, M., *144 Problems of the Austrian–Polish Mathematics Competition 1978–1993*, The Academic Distribution Center, 1994.

41. Kuczma, M., *International Mathematical Olympiads 1986–1999*, Mathematical Association of America, 2003.

42. Larson, L. C., *Problem-Solving Through Problems*, Springer-Verlag, 1983.

43. Lausch, H. *The Asian Pacific Mathematics Olympiad 1989–1993*, Australian Mathematics Trust, 1994.

44. Liu, A., *Chinese Mathematics Competitions and Olympiads 1981–1993*, Australian Mathematics Trust, 1998.

45. Liu, A., *Hungarian Problem Book III*, New Mathematical Library, Vol. 42, Mathematical Association of America, 2001.

46. Lozansky, E.; Rousseau, C. *Winning Solutions*, Springer, 1996.

47. Mitrinovic, D. S.; Pecaric, J. E.; Volonec, V. *Recent Advances in Geometric Inequalities*, Kluwer Academic Publisher, 1989.

48. Mordell, L.J., *Diophantine Equations*, Academic Press, London and New York, 1969.

49. Niven, I., Zuckerman, H.S., Montgomery, H.L., *An Introduction to the Theory of Numbers*, Fifth Edition, John Wiley & Sons, Inc., New York, Chichester, Brisbane, Toronto, Singapore, 1991.

50. Prasolov, V.V., *Problems in Plane Geometry*, Fifth Edition, Moscow textbooks, 2006.

51. Savchev, S.; Andreescu, T. *Mathematical Miniatures*, Anneli Lax New Mathematical Library, Vol. 43, Mathematical Association of America, 2002.

52. Sharygin, I. F., *Problems in Plane Geometry*, Mir, Moscow, 1988.

53. Sharygin, I. F., *Problems in Solid Geometry*, Mir, Moscow, 1986.

54. Shklarsky, D. O; Chentzov, N. N; Yaglom, I. M., *The USSR Olympiad Problem Book*, Freeman, 1962.

55. Slinko, A., *USSR Mathematical Olympiads 1989–1992*, Australian Mathematics Trust, 1997.

56. Szekely, G. J., *Contests in Higher Mathematics*, Springer-Verlag, 1996.

57. Tattersall, J.J., *Elementary Number Theory in Nine Chapters*, Cambridge University Press, 1999.

58. Taylor, P. J., *Tournament of Towns 1980–1984*, Australian Mathematics Trust, 1993.

59. Taylor, P. J., *Tournament of Towns 1984–1989*, Australian Mathematics Trust, 1992.

60. Taylor, P. J., *Tournament of Towns 1989–1993*, Australian Mathematics Trust, 1994.

61. Taylor, P. J.; Storozhev, A., *Tournament of Towns 1993–1997*, Australian Mathematics Trust, 1998.

62. Yaglom, I. M., *Geometric Transformations*, New Mathematical Library, Vol. 8, Random House, 1962.

63. Yaglom, I. M., *Geometric Transformations II*, New Mathematical Library, Vol. 21, Random House, 1968.

64. Yaglom, I. M., *Geometric Transformations III*, New Mathematical Library, Vol. 24, Random House, 1973.

Other Books from XYZ Press

1. Andreescu, T., *105 Algebra Problems from the AwesomeMath Summer Program*, 2013.
2. Andreescu, T.; Kane, J., *Purple Comet Math Meet! - the first ten years*, 2013.
3. Andreescu, T.; Dospinescu, G., *Straight from the Book*, 2012.
4. Andreescu, T.; Boreico, I.; Mushkarov, O.; Nikolov, N., *Topics in Functional Equations*, 2012.
5. Andreescu, T., *Mathematical Reflections - the next two years*, 2012.
6. Andreescu, T., *Mathematical Reflections - the first two years*, 2011.
7. Andreescu, T.; Dospinescu, G., *Problems from the Book*, 2008.

Index

Printed by "Combinatul Poligrafic"
Com. nr. 30754